SpringerBriefs in Applied Sciences and Technology

From the European Space Policy Institute

More information about this series at http://www.springer.com/series/15974

Matteo Tugnoli · Martin Sarret
Marco Aliberti

European Access to Space: Business and Policy Perspectives on Micro Launchers

ESPI
European Space Policy Institute

Springer

Matteo Tugnoli
European Space Policy Institute
Vienna
Austria

Marco Aliberti
European Space Policy Institute
Vienna
Austria

Martin Sarret
European Space Policy Institute
Vienna
Austria

ISSN 2191-530X ISSN 2191-5318 (electronic)
SpringerBriefs in Applied Sciences and Technology
ISSN 2523-8582 ISSN 2523-8590 (electronic)
SpringerBriefs from the European Space Policy Institute
ISBN 978-3-319-78959-0 ISBN 978-3-319-78960-6 (eBook)
https://doi.org/10.1007/978-3-319-78960-6

Library of Congress Control Number: 2018940881

Printed on acid-free paper

This Springer imprint is published by the registered company Springer International Publishing AG part of Springer Nature
The registered company address is: Gewerbestrasse 11, 6330 Cham, Switzerland

Acknowledgements

The authors would like to express their sincere thanks to ESPI Director, Jean-Jacques Tortora, and ESPI Coordinator of Studies, Sebastien Moranta, for their unremitting and helpful support throughout the course of this project. The authors are furthermore grateful to all the experts and stakeholders interviewed under Chatham House Rule for this report, who contributed with outstanding insights and reflections to the research. Finally, the authors would like in particular to thank Nida Baierl, Jérôme Vila, Chris Larmour, Rolf Janovsky, Florian Loire and Maxime Jambon for reviewing the final draft of the report and providing invaluable suggestions and constructive comments.

Contents

About the Authors

Matteo Tugnoli has worked as Research Fellow at the European Space Policy Institute (ESPI) in Vienna, Austria, since January 2015, where he manages and carries out research projects on space policy and strategy. He further holds the ESPI International Secretariat for the European Interparliamentary Space Conference (EISC) since May 2015. Prior to joining ESPI, he worked as Trainee in the Relations with Member States Department, Director's General Cabinet, of the European Space Agency (ESA HQ) in Paris, France. He started his space career as Research Assistant at the Institute for Radioastronomy, National Institute for Astrophysics (INAF) in Bologna, Italy, where he carried out research on diffuse synchrotron radio emission (radio halos and relics) in clusters of galaxies. Mr. Tugnoli has a Bachelor of Science in Astronomy and a Master of Science in Astrophysics and Cosmology, from the University of Bologna, Italy. He also has a Master in Space Policies and Institutions from the Italian Society for International Organizations (SIOI) and Italian Space Agency (ASI) in Rome, Italy.

Martin Sarret is a French researcher seconded by the Centre national des études spatiales (CNES) at the European Space Policy Institute. The current focus of his research is Space Security and transatlantic cooperation in the field. Martin graduated in Modern Languages—English, German, Italian—and Management with a minor in Astronomy. As an undergraduate, he completed two exchange programmes in Germany, one of which was funded by the German Academic Exchange Office. He holds a M.Sc. in Science and Technology Policy from Sussex Business School (Science Policy Research Unit) and is currently completing an M.A. in Strategic Management at the University of Paris Sud. Prior to joining ESPI, he studied Russian at the French National Institute for Oriental Civilizations and Languages (INALCO). He also worked on Launcher Development Policy and interagency cooperation for the Foundation for Strategic Research, a Paris-based think tank specialized in defence matters and the French National Centre for Scientific Research (CNRS) in collaboration with the OECD Space Forum.

Marco Aliberti works as Resident Fellow at the European Space Policy Institute (ESPI) in Vienna, Austria, where he has carried out and published a number of research projects in the areas of access to space and human spaceflight, governance and international relations of space, and Asia's space programmes, particularly those of China, Japan and India. Prior to joining ESPI in October 2012, he held positions consistent with his academic background in East Asian Studies. Mr. Aliberti graduated in Oriental Languages and Cultures at the University of Rome "La Sapienza" and obtained a M.A. in International Relations from the Italian Diplomatic Academy (SIOI) in Rome. He also completed a Master of Advanced Studies in Space Policy and Institutions with the Italian Space Agency (ASI), SIOI and the National Research Council (CNR); Security Studies at the Institute of Global Studies—School of Government in Rome; as well as International Asian Studies at the University of Naples "L'Orientale", with a specialisation in East Asian Relations.

List of Figures

List of Tables

List of Charts

Chapter 1
Introduction

1.1 Background

When observing recent developments in the worldwide space launch sector, one of the most striking features is the dramatic increase in the number of small satellites missions, in particular since 2013. Indeed, in the span of a few years, technological advancements—among other factors—have enabled small spacecraft to achieve a plethora of missions, which only a decade ago were the exclusive domain of much larger satellites. Even more so, small satellites are being employed in an unprecedented way to build and realise large-scale commercial businesses, from Earth remote sensing to communications, on top of burgeoning small-scale scientific and educational missions from universities and amateur groups. Equally important, there is wide consensus that this trend will accelerate the future.

Analysing in greater detail the small satellite launch market yields further insights into how these trends have developed. In fact, the strategy for launching most of these satellites into low Earth orbit (LEO) has been primarily based on opportunity rides alongside main payloads (piggyback) or deployment from resupply capsules and dispensers from the International Space Station (ISS)—in particular for CubeSats and smaller missions. The relatively low prices charged for these launches appealed greatly to small satellite developers, enabling their small, cheap, mass-produced satellites to access space at reasonable prices and to start generating data (i.e. business) for emerging start-ups. This was particularly so for CubeSats deployed for the Planet and Spire constellations.

In a parallel and unprecedented development, a great number of NewSpace launch systems, mostly private-led, are being developed currently, offering a payload capacity that varies from a few dozen to a few hundred kg to LEO. Their stated objective is to capture the demand for dedicated access to space from this booming small satellite market. Indeed, as small spacecraft become increasingly capable,

M. Tugnoli et al., *European Access to Space: Business and Policy Perspectives on Micro Launchers*, SpringerBriefs from the European Space Policy Institute,
https://doi.org/10.1007/978-3-319-78960-6_1

and their potential applications multiply, the limitations of current launch solutions become more evident, especially when compared to the increasing ambitions of small satellite operators.

Demanding satellite operators are becoming increasingly dissatisfied with factors that inherently limit opportunities and, ultimately, adversely affect the *time to mission/market* of their products. These factors include: lack of flexibility in choosing orbital parameters when launching as secondary "guests" of a main payload (whose owner is the deciding customer of the launch service); unsatisfactory availability of launch slots in a global space launch industry that has seen a fairly stable total number of orbital launches over more than a decade; and lastly, launch manifest delays that result from the main customers' conditions over which the smallsat operator has no control.

It is in this evolving context that micro launcher concepts and initiatives have started to proliferate across the globe in recent years. Far from being a new concept (the air-launched Pegasus XL has been active since the early 1990s), these vehicles are today being developed at the nexus of anticipated demand for their services (from small satellite operators), and favourable dynamics (for example in terms of access to venture capital funding), also within the NewSpace paradigm. Their aim is to fill the perceived gaps in the current space transportation offer for smallsats, with targeted and dedicated services—albeit coming at a higher price than piggyback missions.

However, many questions and known unknowns still hover over these ventures, including on the one that has succeeded in achieving its first successful orbital tests. The consensus of all stakeholders (except perhaps micro launcher companies) is that the prospects for this segment of the space transportation industry are still unclear. The variables include the size and shape of the actual smallsat market to be captured, the feasibility of company business cases and the level of competition with current small-to-medium scale rockets. This applies also to the European context, where the commercial and institutional utility of micro launchers is still subject to debate among the different stakeholders.

1.2 Objectives of the Report

This research responds to the clear need for a comprehensive assessment of the commercial and institutional landscape for micro launchers initiatives in the European context. The primary objective of this report is to offer an analysis of the dynamics of the global launch service market associated with small satellites, by placing particular focus on developing trends and the future outlook.

From an economic perspective, the emergence of numerous micro launcher initiatives on top of currently available launch solutions for small satellites raises questions about the specific nature and size of the smallsat market that will—or could—be addressed by such emerging launch services offerings. The second objective of this report is identifying the specific features of this market by characterising customers' expectations and providing a competitive analysis of the different launch solutions for small satellites.

Finally, and from a policy perspective, the report intends to reflect on whether such a market could and should be handled on a purely private basis in the European context or whether it generates important stakes that compel European institutional actors to become more actively involved in this domain.

1.3 Methodology and Structure

This study has been mainly carried out through desk research of publicly available documents, external and internal databases, conference proceedings and other bibliographic sources, spanning both sectorial and general contributions in the area of space transportation and small satellites.

The research has been complemented and strengthened by 16 targeted interviews, conducted under Chatham House Rules, with high-level stakeholders, experts and representatives of the small satellite industry as well as space transportation companies, national and pan-European space agencies, space policy experts, venture capital funds and academia.[1]

The report includes quantitative data and analyses, which—in line with the general criteria proposed by Eisenhardt and Yin—were mainly collected to support the construction of qualitative research. Furthermore, the market analysis has been conducted according to the criteria for business strategy developed by Osterwalder and Pigneur.

It should also be highlighted that while the approach of this study is markedly empirical and its scope global, its underlying purpose is normative in nature and deliberately European in perspective: the study concretely intends to *describe* and *explain* the dynamics of the small satellite and micro launcher sector in order to *recommend* possible policy actions and strategies to European decision-makers in this regard. This coincides with the European Space Policy Institute's mission to provide European stakeholders with informed analyses in the field of space policy and to facilitate the decision-making process in Europe.

This report is structured as follows.

After providing the reader with the background, rationale and objectives of this study in the chapter to hand, Chap. 2 introduces the topic of micro launchers, starting with a historical perspective, and highlighting how these new developments can be viewed as embedded in the currently unfolding NewSpace dynamic. It further describes the main drivers behind contemporary micro launcher initiatives, both from

---

[1]The stakeholders and experts interviewed for this research are representatives of the following companies and institutions: European Space Agency (ESA), Centre National d'Etudes Spatial (CNES), Italian Space Agency (ASI), UK Space Agency (UKSA), National Aeronautics and Space Administration (NASA), ArianeGroup (AG), OHB System (OHB SE), Surrey Satellite Technology (SSTL), Planet, ICEYE, Orbital Express, Vector Space Systems, NanoRacks, George Washington University Space Policy Institute, Bradspace, Seraphim Capital.

a commercial and an institutional perspective. The chapter concludes with a detailed overview of worldwide micro launcher developments, in terms of reference market, funding origin and status. Further considerations are provided regarding expected launch frequencies and prices of some micro launcher projects, as well as launch sites.

Chapter 3 describes and discusses business perspectives for micro launchers in greater detail. After an explanation of the methods employed in the subsequent analysis, the chapter provides a market analysis of the small satellite launch market, including a detailed assessment of the small satellite market segmentation in terms of mission and customer typology, as well as customer expectations. This part is followed by a competitive analysis of current and future launch strategies for small satellites. These analyses lay the groundwork for a benchmark of small satellite launch solutions with regard to customer expectations, which enables the identification of strengths and weakness and, ultimately, the value proposition of micro launchers—and the stakes for micro launcher companies.

The concluding chapter shifts the focus to providing policy considerations that are relevant in the European context. It begins by describing the status of micro launcher initiatives in Europe, from a demand and supply perspective, with the aim of assessing the "worthiness" of this specific kind of space transportation for Europe. Building on this assessment, the chapter subsequently lists several policy considerations to help ensure strong global European strategic positioning in the market segment for launch solutions for small satellites.

Chapter 2
Overview on Micro Launchers

Throughout this report, we adopt the following classification for launch vehicles[1]:

- A sounding rocket is only capable of suborbital flights and cannot deliver payloads to Earth orbit.
- A micro launcher vehicle is able to place less than 500 kg of payload into low Earth orbit (LEO).
- A small-lift launch vehicle (also "small launcher" or "light launcher" throughout this report) is capable of lifting between 500 and 2000 kg of payload into LEO.[2]
- A medium-lift launch vehicle (also "medium launcher") is capable of lifting 2000 to 20,000 kg of payload into LEO.
- A heavy-lift launch vehicle (also "heavy launcher") is capable of lifting 20,000 to 50,000 kg of payload into LEO.
- A super-heavy-lift vehicle is capable of lifting more than 50,000 kg of payload into LEO.

With regard to LEO orbits, it is important to note that pure payload mass alone is not sufficient to characterise the performance of a launch vehicle, as orbit altitude and inclination (as well as launch site latitude) are also key factors that heavily affect the vehicle's performance. That said, it is not in the scope of this report to make a comparative assessment of launch vehicle performance.

2.1 Historical Perspective

Small-scale space launchers are far from being a new concept: historically, every spacefaring nation started developing its own space programmes with rockets whose capacity were akin to what today we would include in the so-called micro launcher

[1] Adapted from McConnaughey et al. (2010).

[2] This class is represented today by vehicles such as VEGA and PSLV.

M. Tugnoli et al., *European Access to Space: Business and Policy Perspectives on Micro Launchers*, SpringerBriefs from the European Space Policy Institute, https://doi.org/10.1007/978-3-319-78960-6_2

category. Building upon the know-how developed in rocketry in the late 1930s and 1940s, then continuing with the likewise emerging ICBM technology,[3] the very first operational launcher of the space era was Sputnik 8K71PS, which in 1957 launched Sputnik 1, an 80 kg payload, into orbit from Baikonur.[4]

A range of small rockets were then developed by the USA., Russia as well as the UK and France, for the twin purposes of achieving suborbital flights and conducting atmospheric research (including, in particular, sounding rockets),[5] as well as to expand and improve national orbital delivery capacities. The American Scout-G1, with its capacity of 220 kg to LEO, was active between 1979 and 1994, when it delivered several experimental and scientific payloads over 18 launches.[6] The British Black Arrow, developed earlier during the 1960s, achieved only a few launches before being retired in favour of the American Scout.

As the USA and the USSR were engaged in the space race, launcher designs moved towards larger vehicles in line with the ambition to place into orbit increasingly heavy and complex spacecraft as well as supporting human spaceflight endeavours, which were even more demanding in terms of performance and safety requirements. While other space powers emerged over the years, their initial launcher designs were quickly scaled upward, as technology improved and payloads mass increased in parallel. Thus, over the first decades of space activities, the development of small launchers (other than sounding rockets, used mainly for scientific research) constituted only but a technological first step towards building larger capacity, as payload size increased quickly and the essential trend was towards an increase of payload mass to orbit. Indeed, as the space era approached maturity, the diversification of launcher capacity was pursued through different categories of launchers.

The first launch of Conestoga 1 in September 1982, built by Space Services Inc., and designed with a modular structure starting with a 500 kg payload capacity into LEO, marked an important milestone in the history of commercial spaceflight: it was the first privately funded commercial rocket that successfully reached orbit.[7] Interestingly, Conestoga 1 as well as the larger Kayser-1 (OTRAG-1)[8] were intended to dramatically reduce the cost of accessing space, through simple designs and a modular/cluster architecture. Together with the establishment of Arianespace as the very first commercial launch service provider in 1980, these events marked the beginning of commercial spaceflight. It is in this historical context that the US Commercial Space Launch Act of 1984 laid the groundwork for regulating US commercial

[3]Some examples are US Vanguard (23 kg to LEO, 1957), UK Black Arrow (132 kg, 1969), India ASLV (150 kg, 1987).

[4]Zak (2017).

[5]A detailed history of sounding rockets, particularly in the European context, can be found in Seibert (2006).

[6]Krebs (2018).

[7]Space Services Inc. of America (2014).

[8]OTRAG was a West Germany private company founded in the late 70s, which developed a fully modular, easy to scale-up launcher with capacities ranging from 1 t to several tonnes to LEO. Its efforts were hampered by missile technology proliferation concerns by other major space powers in the context of the Cold War, and ultimately shut down in the 1980s.

spaceflight through a licensing mechanism, further paving the way for the involvement of private companies in launcher development.[9]

In the 1990s, with ICBM technology reaching maturity and the emergence of the first international treaties and agreements to gradually reduce the superpowers' nuclear arsenals and missiles, both the USA and the USSR (then Russia) started converting their surplus ICBMs to a variety of small-sized orbital launchers. This applies to both the Russian Start-1, which achieved five launches between 1993 and 2006, and the American Minotaur-1 employed between 2000 and 2013.

In the late 1990's, the history of micro launchers was also marked by the emergence of the first large LEO telecommunication constellation projects (i.e. Iridium, Orbcomm and Globalstar).[10] Expectations at that time were that hundreds and hundreds of small satellites would be deployed in orbit to build up an entirely new market, namely mobile telephony services provided by satellites. While most of those payloads were delivered into orbit by medium or large-sized launchers, the smaller satellites of the Orbcomm constellation were deployed by the air-launched micro launcher Pegasus, notably because both solutions had been developed by Orbital Sciences. However, constellation companies, incapable of generating revenue and recovering development costs, and having to compete with the rapid expansion of cheaper, ground-based cellular communication, filed for bankruptcy shortly after their deployment in the early 2000s. This development put a sudden stop to the prospects for further employing small launch vehicles in the deployment and replenishment of such constellations.

Over the 1990s, one of the few, if not the only, micro launcher successfully active in the commercial market remained Orbital ATK's Pegasus and Minotaur-1. It is quite revealing to look in more detail at the typology of their customers. Minotaur-1 achieved 11 successful launches between 2000 and 2013, with a manifest of mixed payloads ranging primarily from defence agencies to university educational projects and technology demonstrators as secondary payloads.[11] During its active history, Pegasus launched over 94 satellites into orbit, two-thirds of which were microsatellites ranging from 18 to 115 kg. It was extensively used also by US agencies such as the Defense Advanced Research Projects Agency (DARPA), Missile Defence Agency (MDA), National Aeronautics and Space Administration (NASA) and US Air Force (USAF), with military missions constituting approximately 15% of the total number of missions, as well as delivering several scientific payloads, also in very recent times. As noted above, the other anchor customer of Pegasus was Orbcomm, which launched 36 satellites of 45 kg each over a decade. In the 90s, Orbcomm booked entire Pegasus launches for cluster launches: one with 8 satellites in 1997, two with 16 satellites in total in 1998 and one with 7 satellites in 1999. Pegasus

[9]Fought (1988).

[10]In the 90s, Globalstar launched 60 satellites weighting 450 kg on Delta, Soyuz and Zenit launchers. Iridium launched 90 satellites weighting 750 kg on Delta, Long March, Proton launchers. Orbcomm launched 38 satellites weighting 45 kg exclusively using Orbital's Pegasus launcher.

[11]Notably, since Minotaur-1 employs surplus ICBM Minuteman II solid rocket motors, US law forbids it to be used for commercial purposes.

also launched a few other constellation satellites such as those for Orbimage (which became GeoEye and was merged in DigitalGlobe in 2013) and Teledesic. Pegasus was mostly launched from Vandenberg, but also Edwards Air Force Base, the Wallops Island and, interestingly, even once in Spain from the Base Aera de Gando launching Instituto Nacional de Técnica Aeroespacial (INTA)'s 200 kg Minisat 1 in 1997.[12] In 2016, the launch vehicle performed its last mission for NASA putting into orbit eight hurricane research microsatellites (Cyclone Global Navigation Satellite System—CYGNSS) weighing 18 kg each. With such a mixed launch log of institutional and commercial payloads (of the 72 satellites launched in the 90s, 28 were launched for US government agencies) and its high operational flexibility, Pegasus successfully delivered small payloads to orbit over the course of two decades, albeit with infrequent flights and a very high price, in particular, in more recent years for NASA missions.[13]

As noted, the development of space launchers over the decades and until the early 2000s saw their performance evolving and increasing in a parallel process with the needs of their customers, both institutional and commercial, and in particular with the growing size of GEO telecommunication satellites, whose design also evolved for better performances, a longer lifespan and different power standards and channels.[14] This iterative process gave birth to a rich diversity of launch vehicle blueprints, exemplified in today's great variety of launch options and configurations in the most advanced spacefaring nations.

On the policy and regulatory side, commercial access to space was also marked by the US 2003 Commercial Space Launch Amendments Act. With this act, the Federal Aviation Administration (FAA) was given the responsibility for steering commercial spaceflight and became "responsible for regulatory oversight of commercial space launches and the granting of operating licences".[15] Of particular importance was the experimental vehicle permit clause that facilitated the approval procedure for test flights of NewSpace vehicles, whereas the licensing regime imposed strict requirements and involved long administrative procedures. It can be argued that this was one of the developments that ultimately paved the way for today's new wave of private-led initiatives in access to space, including micro launcher-based projects, across the USA.

The recent evolution in micro launcher development must be put in context alongside the equally relevant evolution in small satellites technology, a development initiated at the turn of the century that gained pace in the next decade. Once again, technological advances in satellite development and, more precisely, miniaturisation processes that allow for smaller payloads to achieve progressively better performance are correspondingly impacting the historical evolution of launch methods as

[12]Clark (2016).

[13]For CYGNSS, a firm-fixed-price launch service task order contract worth approximately $55 m was concluded between NASA and Orbital ATK. See National Aeronautics and Space Administration (2014).

[14]United Nations (1998).

[15]Hancock (2005).

the increasing need for launch slots and flexibility in orbital deployment is leading to a greater number of launch options.

As dozens of new micro launchers are being developed today, it seems that the space transportation sector is on the edge of a renewed offering of solutions adapted to smaller payloads, led by the private sector and driven by an expected demand growth itself stimulated by new dynamics in the satellite market. Yet, whether these expectations will be fulfilled is inherently connected with and dependent on the shape of the expected booming market of small satellites, competing strategies from established launcher companies and ultimately the price of access to space they can offer.

2.2 A Trend Embedded in the NewSpace Dynamic

The growing excitation to develop micro launcher vehicles in recent years, and associated launch solutions, is embedded in the evolution of the global space sector observed in recent years, and consensually referred to as "NewSpace".[16] While this does not apply to those developments originating from established companies (i.e. scaled-down versions of current small launch vehicles), a description of the NewSpace dynamic can give better understanding of the context behind a variety of private initiatives.

NewSpace is a sectorial dynamic encompassing various interrelated trends which, together, are driving the emergence of a new approach to space activities characterised, in particular, by a more prominent and leading role for the private sector and by more private, market-oriented activity. From this standpoint, NewSpace is often contrasted to "Traditional Space" which is itself characterised by an activity essentially piloted by governments, motivated by strategic and political objectives and highly dependent upon public funding for space systems development, launch and operations. The situation is obviously more complex as NewSpace is the result of both the rise of new business models and of a progressive transformation of the traditional way of conducting space activities. In fact, governments have been instrumental in the emergence of the NewSpace ecosystem through the implementation of various initiatives aimed at fostering the development of private businesses in the sector. These public initiatives include, in particular, evolution of the legal and regulatory frameworks for the conduct of commercial space activities and the implementation of new public procurement schemes removing the prescriptive constraints of traditional cost-plus contracts and building on a more prominent role for, and investment by, private actors. It is actually the successful combination of effective public strategies and favourable business conditions that has led to the trends observed today.

Although there is no broadly accepted definition of NewSpace, a recent study from ESPI identified the related trends (Fig. 2.1).

---

[16]Section 2.2 contains extracts from the report "The Rise of the Private Actor in the Space Sector" published in the "SpringerBriefs from the European Space Policy Institute" series (Vernile, 2018).

Fig. 2.1 Main trends of the NewSpace dynamic. *Source* Vernile (2018)

There is evidence that the space sector is currently undergoing a progressive change in its structure and that the historical status quo for the conduct of space activities is increasingly being challenged. The first sign of this sectorial transformation is the announcement and initial development of numerous business-driven endeavours in different segments of space activity (i.e. launchers development and launch services, Earth observation, telecommunication, space exploration). The vast majority of space activity is still driven by governments with private industries acting as contractors for public programmes and massively relying on public funding. Notwithstanding, the recent ambitious plans announced by a number of commercial companies and backed by private funds as an indication of confidence in the potential profitability of these business plans have put the spotlight on the rise of a more private, market-oriented leadership in the space sector.

These endeavours are principally the outcome of new entrants in the sector including new businesses, start-ups and large ICT companies eager to build on cross-fertilisation with space applications (e.g. Google). These new businesses leverage private and/or public funding to initiate innovative business models and address NewSpace markets or existing space markets with disruptive solutions.

NewSpace endeavours are usually characterised by innovative concepts including:

- **Innovative industrial approach**: industrial organisation optimisation, supply chain rationalisation and vertical integration, miniaturisation, proven technology reuse, economies of scale, production line automation and digitisation, standardised architectures, use of commercial off-the-shelf (COTS) components or under-specification with alternative risk mitigation methods (e.g. through replacement or in-orbit servicing).
- **Market disruption solutions**: development of solutions with the capacity to disrupt existing markets or to create new markets. A few value proposition features can be identified as being typical of NewSpace solutions: aggressive pricing strategy, solution integration/customisation (i.e. adapted turnkey solutions), flexibility, availability (i.e. lead-time reduction) or decomplexification, among others.

In general, NewSpace is also associated with the emergence and growth of new market verticals. In the downstream part of the value chain, various promising new markets have been identified for business ventures including global connectivity, geo-information services, space tourism and, in the longer-term, space mining. The provision of such new services requires specific systems, which, in turn, impact upstream activities with the development of specific solutions. Among the growing upstream verticals, the skyrocketing number of small spacecraft launched into space, including CubeSats or megaconstellations, has created momentum for businesses interested in providing dedicated solutions (e.g. micro launchers, miniaturised satellite systems, COTS). The profitability and sustainability of NewSpace business models still have to be demonstrated, nevertheless, and although the future of NewSpace endeavours remains uncertain today, the development of these projects has already created a new dynamic impacting the overall space sector.

It is in this evolving ecosystem that the increasing number of micro launcher projects must be examined as this trend is fully embedded within the general NewSpace dynamic. From this perspective, micro launcher projects are a direct consequence of the anticipated growing need for adapted solutions resulting from NewSpace business initiatives (i.e. megaconstellations, CubeSats, commercial demand growth) and are, themselves, often characterised by NewSpace features (e.g. entrepreneurship, private investment, innovative industrial approaches, market disruption solutions).

2.3 Drivers Behind Micro Launcher Initiatives

Access to space is the backbone of any space-related programme, either institutional or commercial. For any business endeavour based on physical space infrastructure, the substantial costs and risks associated with launching space systems to orbit have a great influence on the overall conduct of the business itself. As a consequence, the development of adapted launch solutions is both a potential business opportunity, in itself, and an essential enabler for the emergence of a sizeable and dynamic market for space-based services and products.

With regard to micro launchers, several drivers behind the proliferation of companies and projects can be identified. For the vast majority of them, two market-oriented drivers are at the basis of these initiatives: the anticipated business opportunity of the emerging smallsat market, and the favourable business conditions that have framed the NewSpace dynamic of the past decade. However, in some instances, the development of a number of vehicles is notably stimulated by other, institutional-oriented drivers. One of them is the goal of achieving a first-ever domestic access to space capacity, in a technological development process akin to the early space age developments of today's space powers. And finally, in some of the most developed spacefaring nations, a micro launcher vehicle is emerging as a very appealing solution in the context of modern and future military space activities to enhance resilience of critical systems.

The following chapters will detail to a greater extent the drivers behind micro launcher initiatives.

2.3.1 Commercial Drivers

The current plethora of new micro launcher initiatives is gaining increasing traction, in particular led by the private sector. These ongoing developments can be explained by several factors, as follows:

- The expected booming market for small satellites, with expected new commercial launch opportunities to be met by smaller, cheaper and more flexible launch options completing and expanding the existing offer.
- Launcher-related technology and know-how are fairly more accessible today than previously. For example, there are several instances where experienced professionals from established launcher companies have established their own micro launcher company, building on their extensive knowledge as well as adopting for their enterprise innovative engineering approaches made possible by advancements in basic technology and science.
- A large amount of private funding and venture capital is available for entrepreneurs in the NewSpace paradigm, as an indirect result of fiscal policies pursued by major central banks with the objective of countering the financial crisis initiated in 2007 and stimulating the economy of developed countries.
- Increasingly, favourable legislative and regulatory environments, particularly in the USA, for commercial, privately led initiatives in space and access to space, ultimately spurring similar developments around the world.

Anticipated Business Opportunity

The main driver behind most micro launcher initiatives in recent years is undoubtedly the expected boom in the small and very small satellite market segment. For most private companies currently involved in micro launcher projects, the ultimate objective is to build a profitable and sustainable launch service business able to capture a share of the future launch demand for such satellites by offering more flexible and adapted solutions to smallsat operators.

As explained in Sect. 3.2, there is widespread consensus that the adoption and use of small satellite platforms, which has visibly gained pace since 2012, are poised to continue and accelerate. Furthermore, as smallsats become more capable and complex, able to perform a greater range of missions, the limitations of current approaches and strategies to launch smallsats into orbit are becoming more evident. It is in this context that the new micro launcher initiatives are being developed, although it is clear that the development of a new orbital-capable vehicle is not an easy feat: initial development costs are very high, not to mention the technological complexity, coupled with the necessity of achieving economic viability despite very low launch prices (below $10 m per flight) in a future small satellite market whose long-term shape and dynamics are still very unclear.

Favourable Business Conditions

This anticipated business opportunity is further reinforced by particularly favourable business conditions that have characterised the evolution of the space sector in the past decade. This is in particular reflected by a conjunction of factors that range from a positive regulatory environment in the USA towards private space initiatives, to increased technology access and simplification, to the availability of large amounts of private and public funds for start-ups,[17] this latter item also related to the peculiar financial landscape as characterised by extremely low interest rates. Indeed, as emerged from interviews conducted with various stakeholders, it is commonly agreed that the peculiar combination of massive amounts of available venture capital combined with extremely low interest rates is one of the factors contributing to the multiplication of private-led initiatives in the space sector, including and beyond the field of access to space and micro launchers.

Furthermore, a majority of new micro launcher developers are embracing the NewSpace philosophy, particularly in the USA, with the idea of "democratising" access to space and, ultimately, lowering costs of deploying payloads into orbit, which in return could enable substantial growth in launch demand.[18] In the USA, such ventures have hugely benefited from less restrictive regulation on launch activity and the amount of funding made available by investors who have a positive attitude towards risk-taking and strongly believe in the big promises behind digital transformation letting space applications diffuse in the economy. Moreover, the so-called entrepreneurial spirit of Silicon Valley is also very much anchored in these ventures and the financial backers who invest in them. In Europe, such ventures are following similar philosophies, despite having a somewhat more limited access to funding, which might also explain a slower pace of development. However, it is noteworthy

[17] With regard to government technology, seed funding opportunities in the USA, see for example, the Federal Business Opportunities—FedBizOps. Web: https://www.fbo.gov/.

[18] High elasticity of launch service market that remains to be demonstrated, as discussed in the ESPI Report "The European Launchers between Commerce and Geopolitics" (Aliberti & Tugnoli, 2016).

that some European projects are also being developed in a transatlantic setting, i.e. raising funding also from the USA.

2.3.2 Institutional Drivers

As mentioned above, it must be noted that not all micro launcher initiatives currently in development have the smallsat commercial market as their objective. Several projects are indeed backed almost entirely by institutional funding, as emergent spacefaring nations are willing to build their first indigenous launcher vehicle, and ultimately to acquire independent access to space capacities. With a limited space industrial base, micro launchers appear to be an economically viable and technically necessary first step to establishing and later expanding their launcher vehicles. One additional driver is the role that micro launchers could play for military space and national security purposes in the most advanced spacefaring nations, particularly in the USA and China.

Independent Access to Space

The political will to establish an independent way to access space is still a powerful driving force for aspiring spacefaring nations. Countries who intend to autonomously conduct their space activities without relying on foreign launch service providers consider launcher technology as a critical element of their space policies and strategies, as analysed in a previous ESPI report.[19]

It must be noted that since the early stages of the space era, concerns around proliferation of ICBM-enabling technology have led to various efforts, mostly from the U.S., to limit the proliferation of such technology in countries seeking to develop domestic launch capabilities, urging them to redirect their research trajectories towards more civilian uses and to join the Missile Technology Control Regime (MTCR) in exchange of more commercial and bilateral cooperation opportunities.[20] But even now, despite these efforts and aside from the large number of private, market-oriented, private-led initiatives to develop a micro launcher discussed in this report—and obviously with this different scope and purpose—it is worth noting that in several countries the development of a small-sized launcher is directly funded by national governments, as a first step towards achieving independent access to space.

Overall, as is clear from the overview of Sect. 2.4.1, one can see that small launcher development programmes are still of interest for countries wanting to access space independently. However, apprehensions over space proliferation are particularly focused on countries labelled as "states of concern" by the USA, specifically Iran and North Korea. In this sense, the security dimension related to the multiplication of micro launcher initiatives worldwide remains to a large extent an element of quiet apprehension.

---

[19] Aliberti and Tugnoli (2016).

[20] Aldrin (1998).

Military Operational Response Capabilities

Micro launchers could play a significant role in improving military response capabilities. Space assets are critical infrastructure for military operations and national defence (communications, intelligence, early warning, etc.). For this reason, defence actors are seeking capabilities that can quickly deploy space assets to meet a specific need or to quickly replace deployed space assets that have been accidentally or deliberately neutralised. The development of small systems, and by extension, small launchers could play an essential role in achieving such quick-response tactical space-based capabilities.

Micro launchers fit into this picture because they can be quickly deployed and launched (*operational responsiveness*). This applies even more when considered in conjunction with the anticipated increased use of smallsats for military purposes, following the ongoing trend towards increased *resiliency* and *disaggregation* of military space systems.[21]

Indeed, throughout the past decade, defence institutions in most advanced space-faring nations have put the spotlight on these topics, seeking preliminary development of vehicles that can meet these requirements. For some decades, the US Air Force was involved in vehicle development aimed at promoting responsive access to space. This involvement was ultimately thwarted by several factors, such as the perception that this distinctive feature wasn't a "compelling military need", the very high costs of development, and a cumbersome procurement process.[22] The establishment in 2007 of the Operationally Responsive Space (ORS) Office exemplifies the continued will of the USA to explore and address its needs for quick and flexible access to space through rapidly deployable orbital capabilities.

Although in a different context and operating under different boundary constraints, options to meet the need for responsive launch capabilities for military purposes have also been explored in Europe. Preliminary studies, undertaken as early as 2007, in France in particular, were conducted to ascertain the potential role and relevance of microsatellite launch vehicles for institutional and defence customers. Several concepts, such as the Aldebaran testbed study project, were initially developed at the level of national space agencies supported by the European aerospace defence industry. Of note, the Astrium/Dassault Aviation project MLA was an air-launched vehicle able to deliver up to 150 kg into LEO. However, similar to their US counterparts, none of these projects has yet reached a fully operational phase.

Despite these setbacks, the emerging involvement of private micro launcher companies in addressing the needs for quick operational access to space for defence purposes is of particular relevance. To illustrate, in late 2017 Virgin Orbit announced that it had received, from a US Defense Department agency, its first launch contract for a LauncherOne mission. The contract involves the launch of a selection of Space Test Programme payloads, the details of which are as yet undecided.[23]

---

[21] In this regard, a US perspective is offered in Taverney (2017).

[22] Air University (2017).

[23] Foust (2017f).

While much less information is publicly available in this regard, it must be mentioned that China has institutionalised the strategic need for quick access to space by embedding it in its National Space Policy through the ongoing development of Kuaizhou ("Fast Vessel"), by the China Aerospace Science and Industry Corporation, which is said to have "*improved China's emergency response capacity in space*".[24]

2.4 Current Developments

2.4.1 Projects Overview

There is great diversity in the micro launcher projects currently under development. The selection of micro launchers presented in this chapter (and in particular in Table 2.1) includes a list of current and planned initiatives across the world, including, for reference purposes, also those that have been cancelled.[25] This section provides a first-order distinction to help categorise the different projects according to specific criteria.

The vehicles are classified into two categories by their **reference market**:

- *Commercial* indicates that the vehicle is primarily aimed at the worldwide commercial launch service market for small satellites.
- *Institutional* indicates that the project is being developed primarily by national states and public entities for the internal institutional market (including defence), with the purpose of developing access to space capacities, or for technology demonstrator purposes, or for institutional smallsat missions.

Similarly, as regards the **visible financial backing** of the projects:

- *Public* indicates that the venture has received funds from contracts, grants or other types of disclosed funding coming from national space agencies or other public institutions.
- *Private* indicates that financing comes from single individuals, industrial players, venture capital and so on. Specific mention is made when a project has raised money through crowdfunding as well.

The **sources** of funding for launcher companies are difficult to identify. A number of start-ups prefer to operate "under the radar" for the initial phases of development, while projects that exist only at "announcement" level sometimes have not yet raised significant capital or received grants. For this reason, a number of vehicles in the table do not show information about their financial backing.

---

[24] Xinhuanet (2016).

[25] As discussed in the introduction, for the context of this report a micro launcher is defined as an orbital-capable rocket with a payload capacity of approximatively under 500 kg to LEO. All data in Table 2.1 from public available sources and databases, and elaborated by ESPI.

The **status** indicates the estimated level of maturity of the initiatives as of early 2018, with further information where applicable:

- *Announcement/Studies* indicate that the company has outlined its plans to develop a micro launcher vehicle or presented R&D concepts to the public.
- *Ground Tests* indicate that the development has advanced to the stage of conducting ground tests of engines and other elements.
- *Flight Tests*, or prototype testing, indicate that the vehicle is reaching a prototype-level, undergoing flight tests (suborbital and/or orbital, including tests of scaled-down versions of the final vehicle).
- *Operational*.
- *Cancelled*.

Table 2.1 Overview of worldwide micro launcher projects (defined as having a payload capacity up to 500 kg to LEO)

Overview of Worldwide Micro Launcher Projects							
Country	**Launcher Name**	**Company Name**	**LEO Capacity (Kg)**	**Reference Market**	**Scheduled First Flight**	**Visible Financial Backing**	**Status (early 2018)**
Argentina	Tronador II	VENG S.A.	200	Institutional	2020	Public	Flight tests
Australia	AUSROC IV	ASRI	35	Institutional			Cancelled
Australia	AUSROC Nano	ASRI	10	Institutional			Announcement
Australia	Austral Launch Vehicle 2 / 3	Heliaq Advanced Engineering	10-100	Commercial		Private	Announcement
Australia	Eris	Gilmour Space Technologies	380	Commercial	2020	Private	Ground tests
Brazil	VLM 1	IAE	150	Institutional	2019	Public	Ground tests
Brazil	VLS-Alfa	IAE	500	Institutional		Public	Cancelled
Canada	Neutrino I	Open Space Orbital	50	Commercial		Crowd funding	
China	Kaituozhe-1	CASIC	100	Institutional		Public	Cancelled
China	Kuaizhou 1A	CASIC / ExPace	300	Institutional, Commercial	2017	Public	Operational
China	LandSpace-1	LandSpace	400			Private	

(continued)

Table 2.1 (continued)

China	NewLine 1	LinkSpace	200 (SSO)		2020	Private	Ground tests
Europe	SMILE		50 - 115	Institutional, Commercial		Public, Private	Studies
Europe	Sparrow	ArianeGroup	500	Institutional, Commercial		Private	Cancelled
Europe	Q@ts / Q@rk	ArianeGroup	50 / 10	Institutional		Private	Studies
France	Eole / Altair [a]	ONERA (and others)	50 / 150	Institutional		Public	Prototype tests
Indonesia	Pengorbitan 1	LAPLAN/TNI AL	25	Institutional		Public	Announcement
India	Chetak	Bellatrix Aerospace	150 (SSO)	Commercial		Public	
Israel	Black Sparrow (modified)	Rafael Armament Development Authority Ltd	80	Commercial			Cancelled
Iran	Simorgh	ISA	250	Institutional	2017	Public	Orbital test failed in July 2017
Italy	Primo	Leaf Space		Commercial		Private	Announcement
Italy	Quick Launch Vehicle	ELV		Institutional, Commercial		Private	Studies
Italy	Final Launch Vehicle	ELV		Institutional, Commercial		Private	Studies
Japan	Momo	Interstellar Technologies	20 sub orbital	Commercial	2020	Private, Crowd funding	Suborbital test
Japan	SS-520-4	IHI Aerospace	4	Institutional		Public	First successful orbital test in February 2018
Japan		NGSRDP (New Generation Small Rocket Development Planning)	100	Commercial		Public, Private	Announcement
Kazakhstan / Russia	Ishim [a]	Kazkosmos/ Moscow Institute of Thermal Technology	160	Institutional		Public	Cancelled

(continued)

Table 2.1 (continued)

Malaysia	Dedicated Nano Launch Vehicle (DNLV)	Independence X Aerospace	200	Commercial	2022	Private	TRL 2 achieved in 2015
Norway	North Star	Nammo	10	Institutional, Commercial	2020	Public, Private	Ground tests
Portugal		Deimos					Studies
Russia	Taymyr LV family	Lin Industrial	12, 14, 108, 180	Commercial	2020	Public, Private	Flight tests
Spain	Arion 2	PLD Space	150	Commercial	2020	Public, Private	Ground tests
Spain	Bloostar [a]	Zero2infinity	75	Commercial	2019	Private	Prototype tests
Spain	Sagittarius Space Arrow	Celestia Aerospace	16	Commercial		Private	Announcement
Sweden	Rainbow Smallsat Express	Swedish Space Corporation	150	Institutional, Commercial	2020	Public, Private	Announcement
Switzerlan d	SOAR	Swiss Space Systems	250	Commercial		Private	Cancelled
UK	Black Arrow 2	Horizon Space Technologies	350	Commercial			Announcement
UK	Orbital 500R [a]	Orbital Access	500	Commercial	2021	Public, Private	Announcement
UK	Prometheus	SpaceLS	250	Commercial			Announcement
UK		Orbex	220	Commercial		Public, Private	Ground tests
UK	Devon 2	Tranquillity Aerospace	4	Commercial			Announcement
Ukraine	Microspace 2 [a]	Yuzhnoye	40	Commercial			Announcement
U.S.	BOREAS	Stofiel Aerospace	250	Commercial		Private	Announcement
U.S.	CAB-3A [a]	CubeCab	5	Commercial	2018	Private	Announcement
U.S.	GOLauncher 2	Generation Orbit	45	Commercial		Public, Private	Ground tests
U.S.	Haas 2CA	ARCA Space Corporation	100	Commercial	2019	Private	Announcement
U.S.	Intrepid I	Rocket Crafters	250	Commercial	2019	Public, Private	Ground tests

(continued)

Table 2.1 (continued)

U.S.	LauncherOne [a]	Virgin Orbit	500	Commercial	2018	Public, Private	Flight tests
U.S.	M OV	Mishaal Aerospace	454	Commercial			Announcement
U.S.	Electron	Rocket Lab	225	Commercial	2017	Public, Private	Successful orbital test in January 2018
U.S.	Neptune modular rocket family	Interorbital Systems	4.5, 18, 40, 500	Commercial		Public, Private	Ground tests
U.S.	Pegasus XL [a]	Orbital ATK	475	Commercial	1990	Public, Private	Operational
U.S.	Sprite LV family	Scorpius Space Launch Company/ Microcosm	100, 160, 480	Commercial			
U.S.	Spyder	UP Aerospace	7.5	Commercial		Public, Private	Ground tests
U.S.	Starlord	RocketStar	300	Commercial		Private, Crowdfu nding	Announcement
U.S.	SuperStrypi (SPARK)	Hawaii Space Flight Lab/Sandia National Lab		Institutional			Test failed in 2015
U.S.		CloudIX [a]	22	Commercial			Announcement
U.S.		New Ascent		Commercial			Announcement
U.S.		Odyne Space		Commercial			Announcement
U.S.	Terran 1	Relativity Space		Commercial		Private	Announcement
U.S.		Spin Launch		Commercial			Announcement
U.S.	MCLV	Whittinghill Aerospace	23	Commercial		Public, Private	Announcement
U.S.	VALT	VALT Enterprises	22	Commercial		Public, Private	Announcement
U.S.	Vector-R	Vector Space Systems	66	Commercial	2018	Private, Public	Successful suborbital prototype tests in 2017
U.S.	Vector-H	Vector Space Systems	160	Commercial	2020[27]	Private, Public	Announcement
U.S.	Volant	Bspace Corporation	215	Commercial			Announcement
U.S.	Xephyr	Mastens Space Systems		Institutional, Commercial		Public	Cancelled

Projects in an advanced development phase or already operational are highlighted
Source ESPI

Considering that easy and low-cost access to CAD tools combined with the ease of diffusion of PR information allow for very small projects to be widely publicised and to sometimes appear more advanced and concrete than they really are, Table 2.1 highlights projects that have reached an advanced level of development or operational status (in blue). And finally, vehicles that are air-launched are indicated by [a].

As emerges from this overview, contemporary micro launcher ventures are primarily private-led, conducted by start-ups that want to develop their own vehicle to penetrate the existing and upcoming smallsat commercial market, or well-established space launch companies that seek to diversify their offer. Indeed, less than 15% of current micro launcher projects can be characterised as public-led. However, the distinction between private and public applies poorly in certain instances, such as China and Russia. While Chinese space ventures are state owned or at least very much controlled by state corporations (e.g. China Great Wall Industries Corporation—CGWIC—and China Aerospace Science & Industry Corporation—CASIC), Russia's system brings together all existing and traditional space ventures under the aegis of the state owned United Rocket and Space Corporation (URSC).[26] While this represents the current setup for traditional space industry in those countries, it can be observed that more recent small-scale Chinese endeavours such as LandSpace or OneSpace and Shenzhen Yu Long, or Lin Industrial in Russia, may also evolve into a more privately led and commercially oriented approach.

On the public sector side, as described in Sect. 2.3.2, institutional interest and funding in these vehicles respond to three perceived needs: to achieve initial access to space, develop technologies, or achieve quick access to orbit for small payloads. Some such projects are described below.

In South America, Argentina's space agency (CONAE) is actively pursuing the development of Tronador II with a budget of $180 m through 2027. In Brazil, the Department of Aerospace Science and Technology, subordinated to the Brazilian Air Force, has partnered with Germany's Deutschen Zentrums für Luft- und Raumfahrt (DLR) for the development of their VLM-1 space launcher.[27] The Turkish Aerospace Roketsan is planning on going forward with the development of the Turkish Launch System with government financial support. In Iran, the first launch of its Smirogh launcher was conducted in July 2017, which was strongly condemned by the USA through a letter to the UN Security Council as well as new economic sanctions.[28] Furthermore, North Korean ICBM tests in August 2017 reignited the sensitive space proliferation debate.

Projects aiming at achieving quick operational access to space are being primarily developed in the USA. In greater detail, the US DoD has developed a number of projects in this regard, such as DARPA's Airborne Launch Assist Space Access

[26]Zak (2016).

[27]de Selding (2015a).

[28]The USA, joined by Britain, France and Germany, affirmed this test was violating a United Nations Security Council resolution endorsed as part of the 2015 "nuclear deal" with Iran that stated that the country should not engage in activities related to ballistic missile technology which is often strongly associated with space launch for its similarities in propulsion systems.

(ALASA), with contracts awarded to six companies and interrupted in 2015,[29] and Small Air Launch Vehicle to Orbit (SALVO) whose budget is still unclear. Super Strypi was another micro launcher project funded by the ORS Office, but its initial tests resulted in failure in 2015.[30] Notably, in China, the Kuaizhou-1A rocket—developed for quick access to space—achieved a first commercial launch of three smallsats in 2017, having previously flown twice with classified payloads.[31]

The remaining projects, either funded by private money or also backed by governments, are working towards the goal of the commercial market, with a variety of approaches.

A number of space launch start-ups without previous operational experience are developing prototypes and non-orbital vehicles, most often intended as a first step in an incremental development strategy before moving on to orbital-capable designs. For instance, the sounding rockets Nexø I and II of the Danish company, Copenhagen Suborbitals, are being built for technology demonstration in preparation for the much larger Spica launcher. Similarly, the Norwegian company, Nammo, is developing its North Star launcher based on previous designs of the North Star Launch Vehicle (NSLV) sounding rockets. US-based Mastens Space Systems' previous vehicle developments include Xombie, Xoie, Xaero, Xaero-B, Xodiac and Xeus, which are performing suborbital flights.

Micro launcher projects currently under development encompass both ground-to-space and air-to-space vehicle designs. Recalling the overall success of Pegasus by Orbital ATK, companies pursuing air-to-space designs include, for example, American Generation Orbit with its GoLauncher 2, as well as Virgin Orbit with LauncherOne. Among less traditional and more experimental designs, the Spanish Zero2infinity Company can be considered as rather unique as it is in the process of developing balloon-assisted orbital launches with its project, Bloostar.

The diversification of projects further includes an additional distinction between solid-propelled and liquid propulsion (as well as a combination of both). Simple designs tend to favour full solid propellants. Solid propulsion systems are typically cheaper to build and maintain, but their combustion is a one-go procedure, as they cannot be reignited, thereby limiting orbital delivery flexibility. Liquid propellants are also relatively popular even though they typically require more technical knowledge and infrastructure, as they need to be stored in specific conditions. One of the advantages of liquid systems is that they can be reignited, thus enabling the achievement of different orbital insertions with a liquid-propelled upper stage. This option was adopted by Rocket Lab and its Electron, Virgin Orbit and its LauncherOne, as well as American Vector Space Systems' Vector-R.

With the goal of reducing launch costs and prices, innovative manufacturing techniques are a cornerstone element for this wave of access to space start-ups. Additive manufacturing—or 3D printing—is one of the most trendsetting options for micro launcher development, with the potential to greatly reduce costs. Rocket Lab is partly

---

[29] Gruss (2015).

[30] Clark (2015).

[31] Clark (2017).

employing this technology for Electron's key components, while Relativity Space aims at creating entirely 3D-printed rockets.

Lastly, while some companies are focusing on one single design, others plan to develop a whole family of modular launchers or to simultaneously undertake other projects unrelated to launcher technology. This is the case of Lin Industrial, whose Taymir proposed micro launcher family includes four models—Taymyr 1A, Taymyr 1, Taymyr 5 and Taymyr 7—along with other proposed vehicle projects such as Aniva, Aldan, Adler, Vyuga and Zeya. Californian Interorbital Systems' Neptune launcher family includes four designs—N1, N3, N5 and N8 LUNA. The American and Romanian ARCA Space Corporation is involved in lunar lander development along with launcher concepts.

Having already signed a number of launch contracts with a range of customers,[32] a handful of micro launcher companies performed key tests in 2017. Orbital ATK is currently developing an upgraded version of Pegasus expected to be active beyond 2017 and air-launched under an aircraft developed by Stratolaunch. Virgin Orbit tested its vehicle's engines on the ground in February 2017, and Vector Space Systems have tested their launch vehicle in real suborbital conditions. Finally, Rocket Lab achieved its first successful orbital launch in January 2018, deploying three small satellites for Planet and Spire.[33]

2.4.2 Business Considerations

The business case for the development of a micro launch vehicle is by no doubt quite challenging. The cost of developing an operational vehicle can be estimated at least $150 m (estimated cost of development for Falcon 1). Despite the lack of detailed information on development costs and the potential profit-per-launch, by extrapolating from the available data it follows that the return on investment alone cannot be very attractive unless the plan is to launch many dozens of launchers per year.[34] Furthermore, as will be discussed in the market analysis of Chap. 3, there probably is limited room in the market for multiple suppliers in this segment, although geographical and political considerations could probably allow a few providers to survive.

With only one successful orbital launch achieved at the beginning of 2018 by Electron, detailed information about the specifics of their business plans are typically scarce. However, some elements can be outlined based on companies' public announcements and the information emerging from initial launch contracts.

First, it must be noted that in the large majority of private micro launcher projects aimed at the commercial market, a high launch rate is indeed envisioned, and most

---

[32] Some examples are represented by Rocket Lab contract with Spaceflight and Vector's contracts with York Space Systems and with ICEYE.

[33] Foust (2018).

[34] Data collected from internal research and interviews with stakeholders.

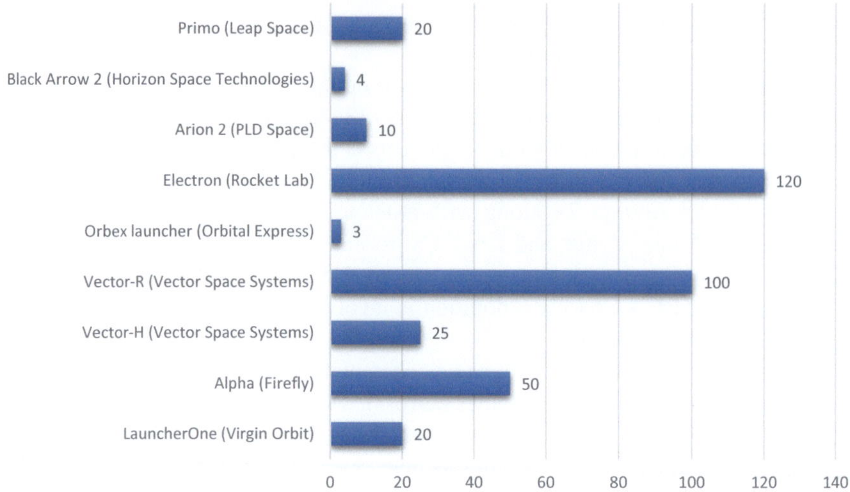

Chart 2.1 Projected full operational launch frequency per year. Cancelled projects are included, for reference. *Source* ESA, Messier (2016), companies' websites

often related to the need to close their business case and cover the very high costs involved in development and subsequent infrastructure investment, enabling a high production rate (such as modern plants and adapted launch pads). Chart 2.1 shows an announced full operational launch frequency per year for some companies.

It emerges that, with a few notable exceptions, for most of the included companies a launch frequency of dozens per year is advertised (probably also to instil investor confidence), notably for all the most mature projects (Electron, Vector-R and LauncherOne). It must be pointed out, however, that all these figures are highly unlikely to be achieved in the medium term. Reaching a cadence of a few dozen launches per year would already be considered an unprecedented feat for any launcher company.

With regard to launch prices, Table 2.2 presents an overview of current estimated price per kg to orbit for selected micro launcher vehicles. The reported price per kg stems from a simple calculation of expected launch price and available capacity and does not consider the concept of filling rate and optimisation of a fairing. Indeed, these cost estimations can be taken as the most optimistic baseline, assuming that the full capacity of the rocket is employed and serve only as a reference for comparison.

As highlighted in the table, the difference in price between the "existing" generation of micro launchers, such as Pegasus XL and the ICBM-derived Minotaur I, and the "new" generation of still-to-be-flown vehicles, is notable. In this regard, the price estimated for Electron, Vector and LauncherOne is significantly lower. How-

Table 2.2 Estimated launch service prices for those micro launcher vehicles whose targeted launch service price is known through company announcements or contracts

Country	Rocket	Performance (Kg)	LEO/SSO	Targeted Launch Service Price ($m)	Estimated Price per Kg ($k)
CN	Landspace-1	400	SSO	7.00	17,500
CN	Kuaizhou-1A	300	SSO	n/a	10,000
IN	Chetak	150	SSO	2.00	13,333
US	SuperStrypi	275	SSO	15.00	54,545
CA	Neutrino I	50	LEO	6.00	120,000
UK	Black Arrow 2	500	LEO	6.60	13,200
MY	DNLV	200	LEO	4.50	22,500
US	Cab-3A	5	LEO	0.25	50,000
US	Firefly Alpha	400	LEO	8.00	20,000
US	GoLauncher 2	45	LEO	2.50	55,556
US	MCLV	23	LEO	2.50	108,696
US	Pegasus XL	475	LEO	56.00	117,895
US	Minotaur I	580	LEO	30.00	51,724
US	Intrepid-1	250	LEO	5.40	21,600
US	Electron	220	LEO	5.30	24,091
US	Vector-R	50	LEO	1.50	30,000
US	Vector-H	110	LEO	3.00	27,273
US	LauncherOne	500	LEO	10.00	20,000
US	Spyder	7	LEO	1.00	142,857

Most advanced or operational vehicles are highlighted. Cancelled projects are included, for reference
Source Catapult Satellite Applications (2017) and companies' websites

ever, their price per kg to orbit is still higher than that offered by other small to large rockets.[35]

Most notably, the Chinese "fast vessel" Kuaizhou-1A advertises a price per kg of only $10k, significantly lower than any other vehicles, and achieved its first successful commercial launch at the beginning of 2017. In this regard, it must be recalled that current export control regulations such as the International Traffic in Arms Regulations (ITAR) strongly limit the presence and access of Chinese launchers on the worldwide market.[36] However, as small satellites are often built with OTC components, the restriction imposed by current ITAR regulations might affect the Chinese micro launch offer in a much less strong manner than it does in the traditional launch market. Furthermore, as Chinese commercial launch operators are increas-

[35] In 2017, estimated prices per kg to orbit for some existing small to large launchers were reported to be the following: Epsilon $40k; Vega, $20k; GSLV, $9400; Proton M, $2800; Falcon 9, $2700. See United States Government Accountability Office (2017).

[36] For a discussion of Chinese launchers and ITAR regulations, see Aliberti and Tugnoli (2016).

ingly active on the global market and offer substantially competitive solutions, the number of ITAR-free small satellite payloads could even increase in the future.

2.4.3 Launch Sites

Any orbital launch activity requires a solid ground segment, including a set of facilities on the launch site to prepare the launcher and process its payload, but also relevant means to follow-up the launch, such as a network of ground stations or emerging new concepts such as flying monitor and control stations as well satellite communication infrastructure. While some micro launchers envisage the possibility of being launched by mobile platforms, this chapter will focus primarily on the topic of spaceports, in particular with regard to the proliferation of new launch sites. Indeed, the underlying assumption is that launch activity will soon rapidly intensify, putting emphasis on the need to have the right infrastructure to support increased launch rates from small and very small rockets.

Several key factors apply to orbital launch sites. The first and foremost is that the geographical location of a spaceport is of prime importance for accessing space and has profound implications on the effectiveness of a launch and its parameters. While equatorial located spaceports are the most advantageous locations for GEO satellites, launch infrastructures situated towards high latitudes can be more convenient for launching spacecraft into polar orbits.

Furthermore, the location of the launch site is inherently connected to safety and security issues. Efforts are maximised to avoid that launch trajectories overpass populated territories, to minimise the risk of damage to humans and properties in case of accident. On top of this, security concerns prompt several spacefaring nations to avoid launching their payloads over the territory of unfriendly countries (such as Shavit from Israel). A third relevant consideration for micro launcher developers is represented by logistics considerations such as, in particular, accessibility for the launch service customer and the distance between the manufacturing plant, test and launch site of the vehicle.

As dozens of micro launcher initiatives are under development, it goes without saying that those vehicles that will reach maturity will necessarily need a location to be launched from. While small rockets could, in principle, be launched even from mobile platforms, in fact an increasing number of new spaceports are being considered and planned all around the globe, on top of plans to accommodate micro launchers in existing infrastructures.[37]

For new commercial spaceports, in most cases two lines of business are being promoted: the professed upcoming space tourism market and small satellites launch opportunities. However, as pioneering private US spaceports have shown, the cost of operating such infrastructures is very high, and the economic returns behind the

[37]Air-launched micro launchers can be considered more flexible in this regard, as they can—in principle—be operated from any kind of suitable airstrip.

market segments they are built for are still unclear. Furthermore, regarding the legal framework of commercial spaceports, it is worth noting that in some countries the demand forecasts for such services and expectations on return on investment are playing a key role in the promotion of such launch infrastructures and are pushing governments to consider enacting appropriate laws to accommodate such businesses. In the USA, 11 spaceports have received their FAA certification, competing by taking advantageous financial measures to attract businesses. These private spaceports include California Spaceport, Mojave Air and Space Port, Pacific Spaceport Complex Alaska and the Ellington Airport. Their offer is mostly about payload processing procedures, maintenance services and security services. Moreover, the micro launcher company Vector Space Systems is seeking to licence the Camden Spaceport in South Georgia[38] to launch its rocket. Spaceport America in New Mexico is an example of a NewSpace infrastructure that as of today seems to be promoting both space tourism and smallsat launch at the same time, combining the two yet different segments in an effort to complete its business case.

In Europe, several new projects are also being considered. In the UK, a number of locations in Scotland, Wales and Cornwall are being assessed as possible sites for a new spaceport. In Scandinavia, the existing facilities at Andøya (Norway) and Kiruna (Sweden), currently in use for sounding rockets experiments, are being actively promoted as potential future small spaceports. In southern Europe, a few projects have been announced, such as the Atlantic Spaceport in the Azores (Portugal) that envisages a €300 m investment plan. A memorandum of understanding was further signed in 2016 between ASI, the Italian national civil aviation authority (ENAC) and the FAA for a possible site in Italy.[39] The same year, an agreement was made between the Italian Altec, Virgin Galactic and Thales Alenia Space to seek ways of accommodating suborbital flights for space tourism opportunities on an existing airport.

The UK is one of the most active countries in Europe when it comes to micro launcher developments and relative ground infrastructure. The UK Space Launch Initiative seeks to set up a domestic launch capability on British soil by 2020. In terms of spaceport infrastructures, a consortium that includes Lockheed Martin UK recently submitted a proposal for a Scottish spaceport, while Orbital Access reportedly partnered with Spaceport Cornwall to become its main operator with their air-launched payload system concept. These agreements and proposals are still at conceptual level and the reality of these business cases is yet to be further investigated as spaceport projects will necessitate private funding to materialise. The government has introduced a seed funding scheme of €11 m to stimulate commercial spaceflight undertakings but will not cover the full development costs. Nevertheless, as part of the UK Space Launch Initiative, it is proactively supporting spaceflight by promoting it as part of its national infrastructure policy goals with active research on regulatory hurdles in addition to seed funding.

---

[38]Foust (2017e).

[39]As announced in a press release by the Italian Space Agency. See Agenzia Spaziale Italiana (2016).

The Norwegian Space Policy encourages the development of comprehensive spaceflight capabilities, by including the development of a launch vehicle, the North Star, further developing the Andøya Space Centre as national spaceport and launching national small satellites for monitoring Norway's territory, which includes swaths of the environmentally sensitive Artic region, as well as ever-increasing maritime traffic on polar routes. Andøya is also the selected base of operation for the SMILE programme, a multinational micro launcher study supported by the European Commission under Horizon 2020.[40] In this sense, the Norwegian model is of interest as its space policy model encompasses the promotion of autonomous access to space capabilities through a small launcher vehicle and its spaceports, as well as deploying remote sensing satellites of strategic importance for the country.

Owing to its already well-established sounding rocket infrastructures and services, in recent years Sweden has been seeking to expand its offer in terms of smallsat launch opportunities as well as space tourism.[41] In particular, the Esrange Space Center near Kiruna in northern Sweden has been extensively used for testing sounding rockets and includes other touristic facilities. The Swedish Space Corporation, which operates Esrange, is also reportedly developing the SmallSat Express concept, aiming at achieving a launch rate of four times per year. A 2015 report commissioned by the Swedish Government recommended exploring the possibility of expanding and upgrading the Esrange Space Center, in the light of the rapid development in the global sector of small satellites and dedicated launches. Furthermore, in late 2017, the Swedish Government announced it had commissioned a feasibility study on the possibility of adapting the country's Esrange Space Center to launch small satellites.[42]

---

[40]European Commission (2016). Horizon 2020 programme funding: approx. €4 m.

[41]Nordic Investment Bank (2015).

[42]Adamowski (2017).

Chapter 3
Business Perspectives for Micro Launchers

3.1 Methodological Approach

To identify and examine key stakes for emerging micro launcher-based services, various elements should be scrutinised including business models, the competitive landscape, prospective markets and technology and regulatory trends, for example. Among available analysis tools, the increasingly used Business Model Generation and Value Proposition Design frameworks, which comprise a set of models developed by Alexander Osterwalder and Yves Pigneur for business strategy, provide a particularly comprehensive and robust methodological basis addressing a wide range of business components.

The Business Model Generation and its complementary Value Proposition Design are essentially tools developed to support the strategic thinking of commercial companies by providing simple, yet complete, models to catalogue and organise external and internal business factors into a simple, coherent and accurate representation. The methodological framework includes three levels of analysis organised around the concept of business model, represented by the business model canvas below, which identifies nine key components of any business organisation addressing a market (Fig. 3.1).

This business model canvas corresponds to the first level of analysis and provides a structure to examine key business components:

- Key partners for the development of the business including suppliers and other business partners;
- Key activities and resources required to provide the services and products on which the value proposition is built;
- The value proposition(s) of the company identifying how the products and services offered by the business create value for the customers;
- Customer relationships and distribution channels modelling the physical and immaterial relations between the business and its customers;

M. Tugnoli et al., *European Access to Space: Business and Policy Perspectives on Micro Launchers*, SpringerBriefs from the European Space Policy Institute,
https://doi.org/10.1007/978-3-319-78960-6_3

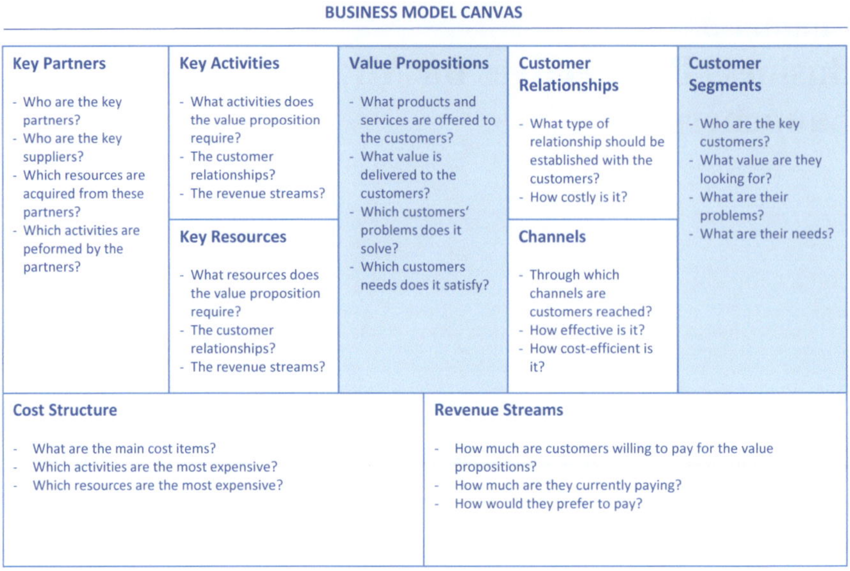

Fig. 3.1 Business model canvas. *Source* Strategyzer (https://strategyzer.com)

- Customer segments identifying target customers and examining how the demand is organised;
- Cost structure and revenue streams identifying the main cost items to provide the value proposition and the way the business generates revenues.

In the frame of this research work, the study team did not attempt to elaborate a generic business model for micro launcher businesses because of the wide range of diverse structures utilised by the different companies and the inherent difficulty of accessing the highly strategic corporate information required to elaborate this model.

The analysis rather focused on the two other levels of analysis of the methodological framework, namely the value proposition and business environment analysis.

The value proposition, which is defined as the value of services and products offered by a business to its customers, is represented by the value proposition canvas below:

As shown in Fig. 3.2, the model focuses in more detail on the description of two central components of the business model canvas and on their interaction:

- the value proposition, which shows how products and services offered by a business create value for its customers;
- the customer segments that identify target customers and their needs;
- the fit between these two components.

The value proposition canvas enables examination and understanding of how products and services offered by a business can fulfil the needs of its target customers.

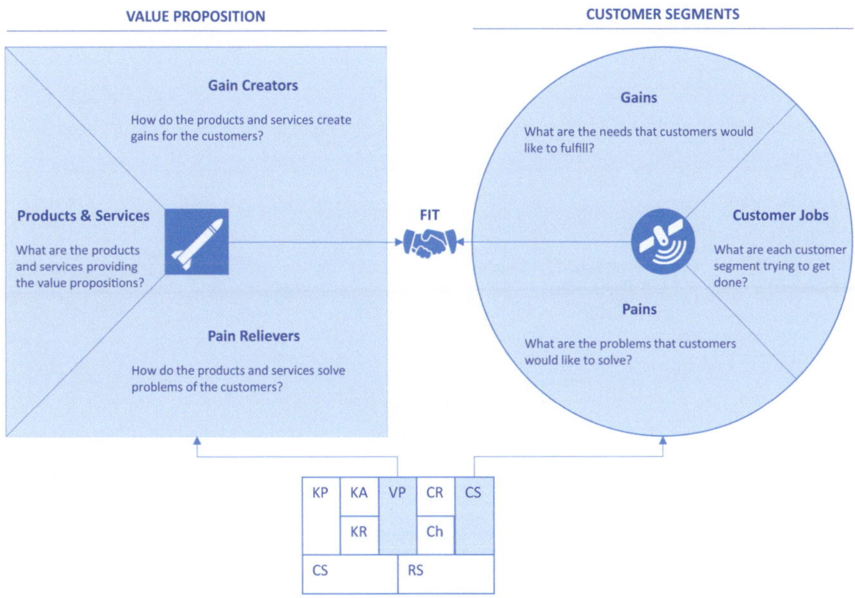

Fig. 3.2 Value proposition canvas. *Source* Strategyzer (https://strategyzer.com)

In the specific case of micro launchers, such analysis enables the description of the current needs of target customer segments and allows the examination of how micro launcher-based services provide, or not, a solution to these needs.

The business environment, which is portrayed in the business environment canvas below, addresses the context in which a business is developed.

As shown in Fig. 3.3, the model organises the different external forces impacting the business. According to the model, the business environment is comprised of four areas:

- Market Forces: Key customers, market segments, market evolution, customer switching costs, etc.;
- Key Trends: Technology innovations, regulatory constraints, etc.;
- Industry Forces: Key actors in the arena such as competitors, new entrants, key suppliers, etc.;
- Macroeconomic Forces: Global market conditions; access to resources; high or low commodities prices, etc.;

In the frame of this study, this analysis has focused on market forces and competitive forces. The objective pursued was twofold:

- Assess the potential market for micro launcher solutions
- Understand how these businesses would be positioned within the competitive landscape.

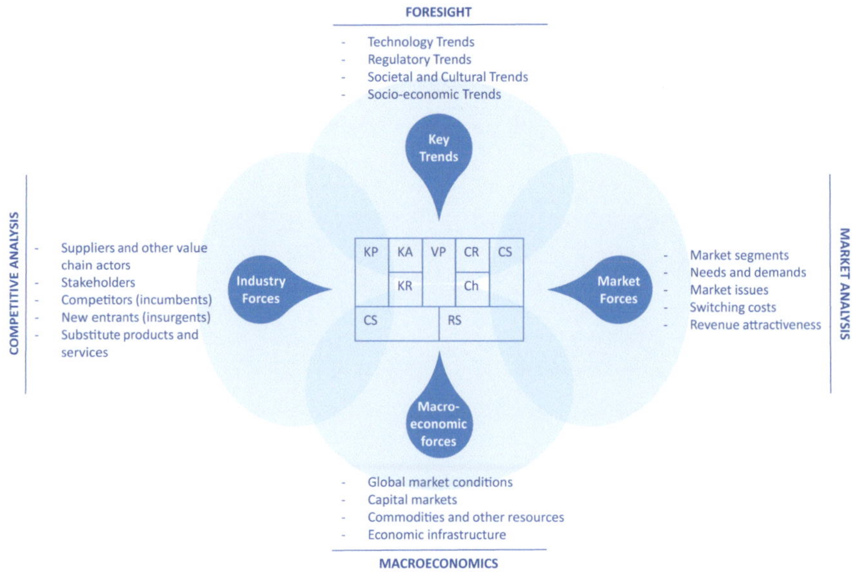

Fig. 3.3 Business environment canvas. Ibid.

Complementarily, the study addresses, in a qualitative fashion, other main elements of the micro launcher business environment and in particular technology and regulatory trends and macroeconomics, such as capital availability, affecting the emergence of these projects.

In sum, the Business Model Generation and Value Proposition Design framework were used in the frame of this study to organise the research work on micro launcher stakes according to an appropriate, robust and widely accepted analysis framework. The study team used this framework to:

- Categorise relevant areas of investigation;
- Conduct an analysis of different internal and external components of micro launcher businesses;
- Consolidate research and analysis results into a coherent set of information and data;
- Identify the main stakes of micro launcher businesses according to the results.

At this stage, it is important to note that, given the wide range of different micro launcher solutions and business models, the analysis performed in this study is based on a selection of projects for which information and data are available publicly or during interviews and which, together, provide a representative sample of current micro launcher projects.

The analysis focuses on generic business elements and on commonalities identified between the different business models. As a result, the conclusions of the analysis

Table 3.1 Classification of spacecraft based on their mass

Classification of small satellites (mass under 500 kg)	
Minisatellite	100–500 kg
Microsatellite	10–100 kg
Nanosatellite	Less than 10 kg

Authors' adaptation from National Aeronautics and Space Administration (2015c) and Catapult Satellite Applications (2017)

may not be valid for all micro launcher projects individually but rather provide a general assessment of the main stakes for this emerging segment of the space launch services sector.

3.2 Market Analysis and Foresight: Key Trends in the Small Satellites Launch Market

3.2.1 Small Satellites Market Trends

In recent years, the global space industry has manifested a significant increase in interest and adoption of small satellites below 500 kg. Within the small satellite category, defined for the purpose of this report as all spacecraft with a launch mass below 500 kg, various subcategories can be identified (Table 3.1)[1].

To match the scope of this analysis, the category "Minisatellite" was expanded to include satellites up to 500 kg. The category "Nanosatellite" was also expanded to include Pico and Femtosatellites.

A historical perspective on the number of small satellites launched over the past decade clearly highlights substantial growth in the segment (see Chart 3.1). Between 2000 and 2012, the total number of small satellites launched in orbit showed a modest increase: until 2012, 37 small satellites were launched on average per year representing between 30 and 40% of spacecraft launched worldwide. The number of these satellites launched into orbit massively increased from 2013 onwards,[2] skyrocketing to more than 130 per year with a peak reached in 2017 with close to 500 small satellites launched in a single year.

This outstanding growth concerned first and foremost the "nanosatellite" category, in particular because of the deployment of a very high number of CubeSats. This specific class of small satellites is based on a standard architecture originally developed in 1999 by California Polytechnic State University at San Luis Obispo (Cal Poly) and Stanford University to provide a simple platform for education to promote and develop competencies to design, manufacture and test space systems.

[1] Adapted from National Aeronautics and Space Administration (2015c).

[2] Primarily owing to the deployment of the constellation of Planet.

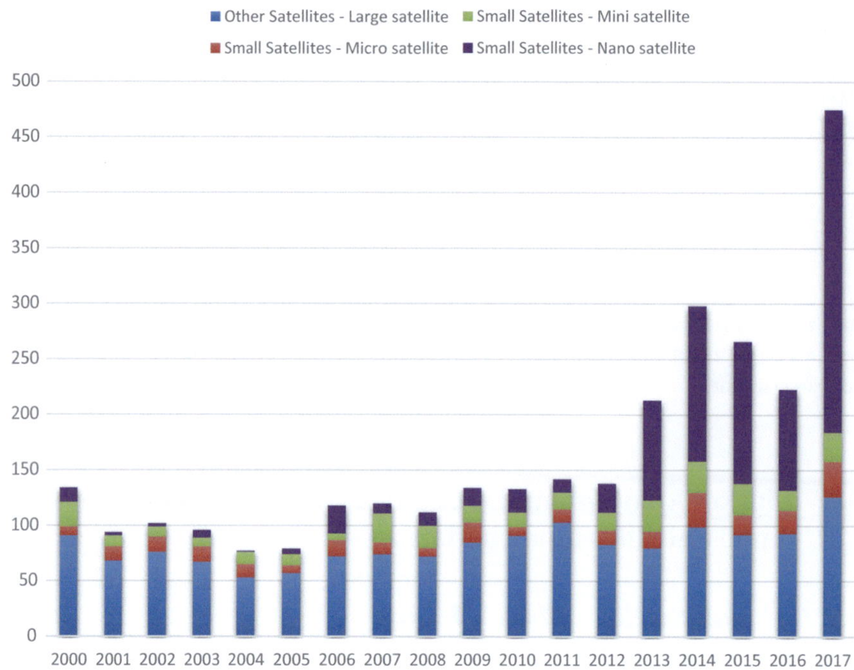

Chart 3.1 Number of satellites launched per year by mass category 2000–2017. For the purpose of this chart, "large satellites" refer to any spacecraft with mass above 500 kg. *Source* ESPI

The CubeSat architecture is based on Units (U) measuring 10 × 10 × 10 cm and weighing around 1.3 kg that can be extended to larger sizes: 1.5, 2, 3, 6 and even 12U. Consequently, a vast majority of CubeSats are included in the nanosatellite category. CubeSats provide a low-cost platform with multiple applications including education projects, scientific missions, technology demonstration and even advanced concepts using constellations or swarms of disaggregated systems. As a result, the standard architecture has met great success since the launch of the first CubeSat in 2003 and has led to the emergence of a specific industry and the development of ever-increasing capabilities for governments, military, commercial operators and academia. Between 2013 and 2017, CubeSats accounted for 71% of all small satellites launched (see Chart 3.2). Between 2014 and 2017, the mini- and microsatellite segment also experienced a slight increase in number, but nowhere near the increase in the nanosatellite segment.

Market figures underline the diversity of applications of small satellites across all "classic" space domains including remote sensing, telecommunication, science, research and technology demonstration and defence (see Chart 3.3). Today, small satellite systems provide a broad range of capabilities ranging from high- and low-resolution spectral and multispectral imagery, weather data collection, narrowband telecommunications, scientific and medical research, and even more. This broad-

Chart 3.2 Share of CubeSats in all small satellites (under 500 kg) launched between 2013 and 2017. *Source* ESPI

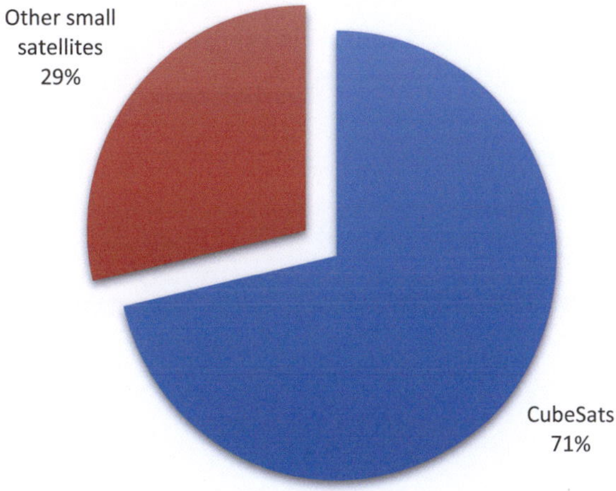

ening is driven by the emergence of new mission concepts and by major improvements in satellite technology miniaturisation both for platforms and payloads. This fast-paced technological evolution has significantly increased the robustness and performance of small satellites (lifetime, altitude and orbit control, capacity and reliability), which have opened new possibilities for this category of spacecraft that was originally used to a limited extent and for specific and usually non-operational missions. Still, providing a low-cost and simpler platform, small satellites and in particular nanosatellites, is massively used for technology and demonstration missions as part of educational, governmental or commercial development programmes. The noticeable growth of the use of small satellites for remote sensing is primarily related to the launch of the Planet constellation of 3U CubeSats (Flocks) providing near-real-time low-resolution imagery, but also to a "sensor revolution" which has substantially increased the capabilities that small satellites can provide in the field of Earth observation.

Small satellite projects have attracted growing interest from a variety of traditional and new customers and investors, including space agencies, military, civil institutions, private businesses and universities (see Chart 3.4). In this field, the US leads with more than 50% of small satellites launched over the last 5 years owned and operated by US-based organisations. China has also developed its use of such satellites and accounted for 15% of small satellites launched in 2015 and 2017. Other space powers, including Japan, Europe, Russia and India, have a more erratic, although non-negligible, use of small satellites. Last but not least, small satellites also represent more than half of the spacecraft launched for countries with limited space capabilities. For these countries, such light platforms provide a relevant solution to get hands-on experience in space systems development and/or to procure space-based capabilities at a lower cost.

Interestingly, the expansion of applications and improved performance has led to a substantial growth of the use of these satellites in the commercial market, notably because they are expected to enable analytics companies to thrive in the billions of

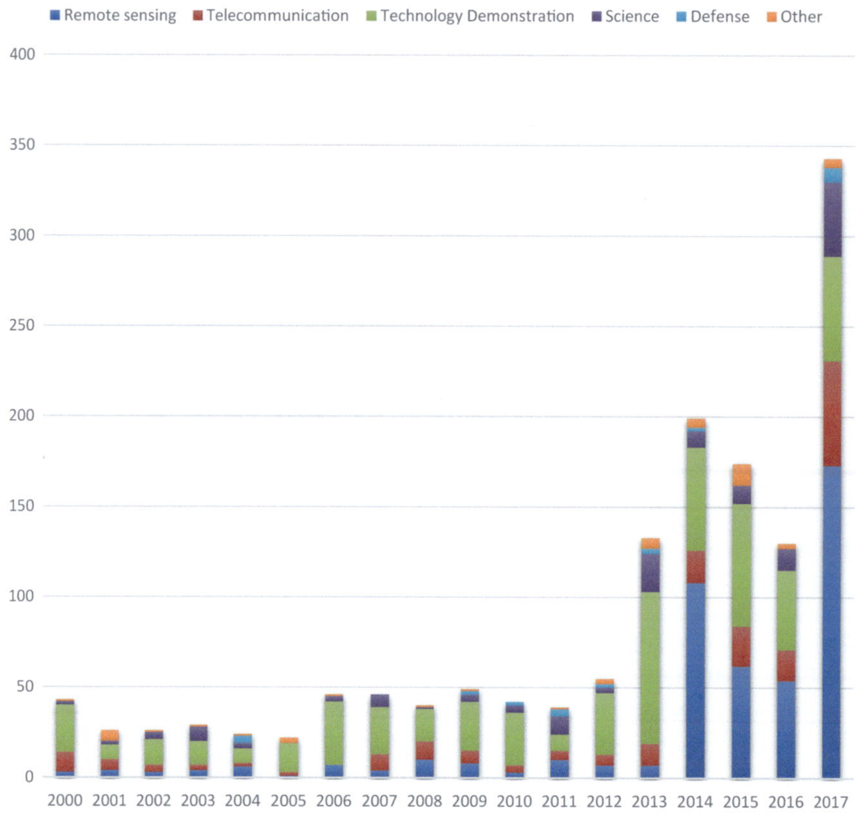

Chart 3.3 Number of small satellites launched by mission type (2000–2017). *Source* ESPI

dollars worth big data market with disruptive space-based capabilities.[3] Although Planet accounts for a majority of these commercial small satellites, more than 30 private companies with different profiles and businesses have deployed at least one small satellite over the last three years: exactEarth, Orbcomm, Spire, Dauria Aerospace, Planetary Resources, Satellogic, Urthecast, Skybox Imaging (acquired by Planet in 2017) to name a few. For these companies, small satellites are either used as development platforms to test and demonstrate their technologies (e.g. electronic components, micropropulsion) or directly to provide services based on capabilities offered by these systems (e.g. imagery, narrowband communications, AIS).

In general, the booming trends observed recently in the small satellite market are the result of multiple factors, among which can be singled out the following main ones[4]:

---

[3]Foust (2016a).

[4]Lim (2016).

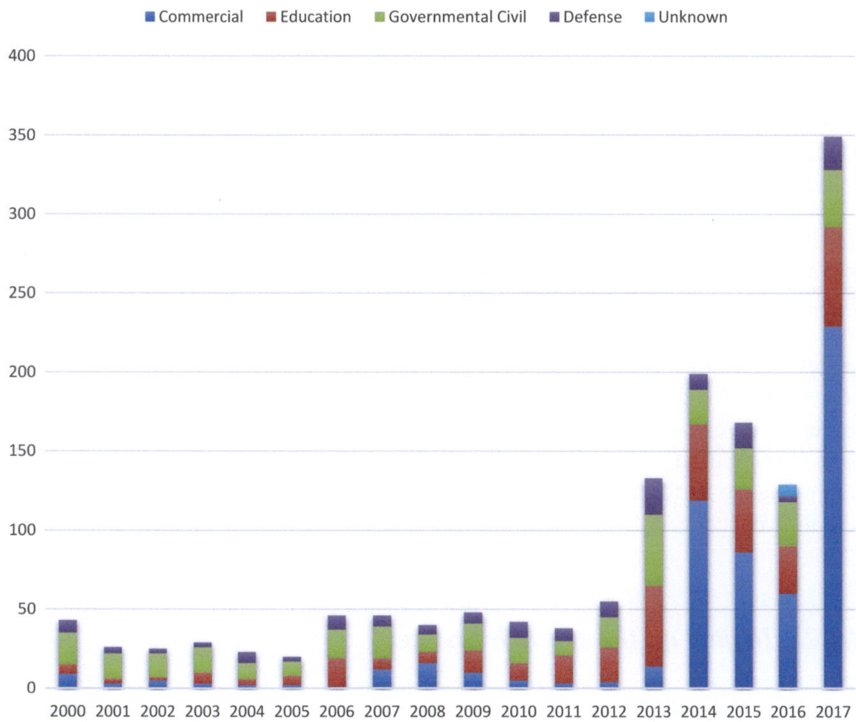

Chart 3.4 Number of small satellites launched by customer nature (2000–2017). *Source* ESPI

- **Standardised architectures**: The development of standardised architectures, in particular of the CubeSat standard, contributed to lowering financial and technical barriers to develop space systems, which triggered a growing interest from universities and start-ups in small satellites.
- **Advancements in satellite technology**: Standardised architectures, miniaturisation of hardware, use of COTS components and new system concepts now enable small satellites to provide a broad spectrum of capabilities, from operational telecommunication and remote sensing services to advanced science missions, and to offer improved capabilities.
- **Low-cost solutions**: The 2008 financial crisis spurred public and private stakeholders to consider with renewed interest more effective and economic methods to build, launch and operate satellite systems. In this context, the affordable costs and improving performances of small satellite platforms constituted a relevant convergence of intents.
- **Digital revolution and NewSpace**: The digital revolution and so-called NewSpace sectoral dynamic created a fertile ground for the development of small satellites through an increased market interest and appetite for global interconnectivity

solutions and commercialisation of big data—particularly in Earth observation, as uniquely provided by innovative space-based solutions.

- **Emerging space-faring nations**: Small satellites provide relevant solutions for countries with a limited space industry to develop domestic competencies for the development and manufacture of satellites and/or to procure space-based capabilities at a lower cost.

These combining factors lie at the foundation of the small satellite growth trend over recent years. This category of spacecraft now enables new companies, including in particular SMEs and start-ups, but also civil and military agencies from all countries, to conduct NewSpace activities and develop innovative business and mission models. From this perspective, small satellites have contributed to lowering barriers of entry in the space sector and to open up a sector that was, until recently, an exclusive domain of established space players.[5] Recent technology developments have also contributed significantly to the increasing adoption of small satellites by improving the performance of these systems and enabling advanced operational and scientific concepts.

3.2.2 Impact on the Launch Market

Despite an indisputable increase in the number of small satellites launched worldwide, this category still represents only a very tiny fraction of the mass—and value—launched to orbit. To illustrate, it takes 6000 1U CubeSats to match just one typical six-ton GEO telecom satellite. When looking at peak years (2013–2017), small satellites represented a mere 2% of the launch activity in mass (see Chart 3.5). This is due to a growth trend having impacted predominantly on the lower end of the small satellite mass category (i.e. nanosatellites and in particular CubeSats). As a result, while the number of small satellites launched per year was multiplied by three to four between the first and second decades of the new millennium, the total mass of these satellites, which was already extremely low, was "only" multiplied by two.

As a reference, in 2014, the 199 small satellites launched to orbit accounted for 8400 kg, mostly launched to LEO. This corresponds roughly to a single launch of a medium-lift launcher such as Soyuz 2.1b. However, these spacecraft were obviously not launched in a single time and around 25% of the launches that occurred that year carried at least one of these small satellites. Of the 92 space launches in 2014, nine carried exclusively small satellites (10%) and 17 carried a mix of small and larger satellites (18%) either piggybacked on primary payloads, or released from the CubeSat dispenser on the ISS, among other solutions (see Chart 3.6).[6]

[5]Rycroft and Crosby (2002).

[6]Platzer (2016).

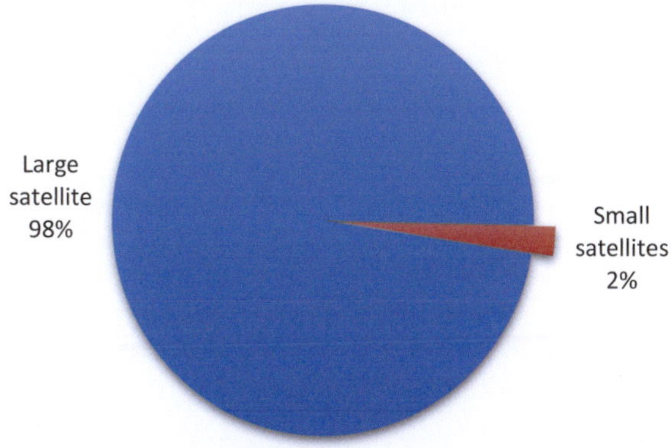

Chart 3.5 Share of small satellite mass (under 500 kg) in the total launched mass between 2013 and 2017. *Source* ESPI

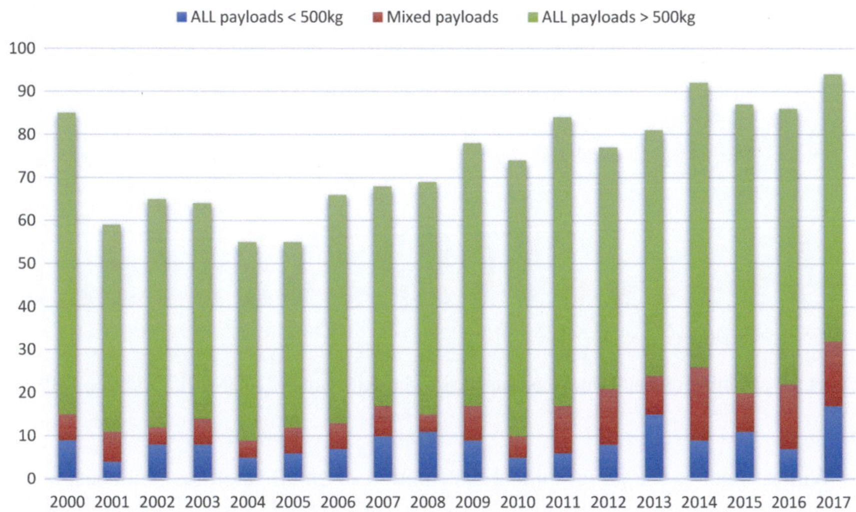

Chart 3.6 Number of launches per year by category of launched payloads, differentiating between those launches which included only small satellites, only large satellites or a mixture of both (2000–2017). *Source* ESPI

A high number of different launchers encompassing light- to heavy-lift vehicles were used: since 2013, 25 different vehicle versions have been used for launches with exclusively small satellites (Rokot-KM, Vega, Minotaur-1, Pegasus XL, CZ-11, KZ-1, Epsilon, etc.) and 26 for launches with mixed payloads (Antares, Atlas-5, H-2A and B, CZ-4B and C, Soyuz-2 in different versions, etc.). Interestingly, a number of launcher versions were used for both launch configurations (Falcon 9, Dnepr, PSLV-CA and XL, CZ-2D).

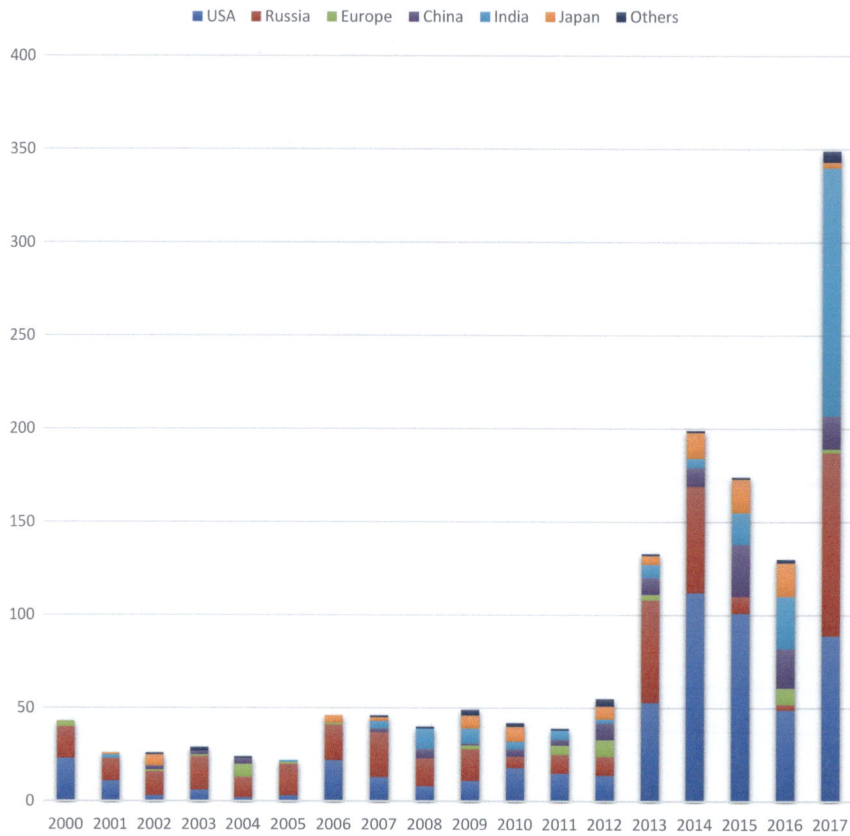

Chart 3.7 Number of small satellites launched by launch country (2000–2017). *Source* ESPI

With respect to the geographical distribution of small satellites launches, at the beginning of the boom in 2013 and 2014, Russia and the US were the orbital launch leaders for small payloads. However, the distribution has been changing considerably in the past three years, with Indian vehicles, for instance, launching around 40% of all small payloads in 2017 (see Chart 3.7).

The launch market also appears to follow the same profile as for larger spacecraft giving way to rather captive markets with a vast majority of small satellites launched with domestic capabilities. Between 2013 and 2017 Chinese, Russian and Indian organisations used exclusively domestic launchers to put their payloads in orbit and 80% of the small satellites of Japanese and US-based organisations used domestic launchers. Europe is the only space power launching most of its small satellites with foreign capabilities (75%) (see Sect. 4.1.2).

To conclude, a few takeaways can be underlined:

- **High numbers but low mass**: The substantial growth of small satellites launched to orbit since 2013 does not fully translate in terms of mass because the growth affected predominantly the lower end of the small satellite mass category. The launch market represented between 5 and 9 tons of payloads launched mostly to LEO.
- **Various solutions**: The launch of small satellites involved a high number of different vehicles in different configurations. The number of launches accommodating at least one small satellite increased, but the number of launches exclusively carrying satellites below 500 kg did not grow substantially. However, recent developments such as the decision by Antrix to set aside 1-2 launches per year for rideshare can be considered a relevant development in the light of their previous approach that centred on piggyback solutions.
- **Captive markets**: As has emerged from the analysis, it turns out that, in line with the global launch activity profile, small satellites are predominantly launched with domestic capabilities, with Europe being a notable exception with almost no captive domestic market in this market segment (see Sect. 4.1.2).

3.2.3 Small Satellites Market Forecast

Although significant growth of small satellites launched per year, in particular for the nanosatellite subcategory, has been observed over recent years, it does not explain, alone, the emergence of such a massive number of micro launcher projects to offer dedicated launch solutions. As a matter of fact, the rationale seems to be more forward looking and related to huge expectations about the future of the small satellite segment. Indeed, available forecasts quite consensually anticipate that the growth recently observed is the early sign of a new massive market.

Among major developments anticipated in the next decade, the return of large and very large satellite constellations composed, this time, of a myriad of small satellites and mostly dedicated to commercial telecommunications and Earth observation applications, appears to be the principal driver expected to maintain and potentially even amplify growth, in terms of both the number and mass of small satellites to be launched. These new large and very large smallsat constellations projects began to draw international attention when, at the end of 2014, a number of applications for large-scale constellations made of hundreds to thousands of satellites were filed at the International Telecommunication Union (ITU).[7] As a basis for comparison, in 2017, the total number of operational satellites in orbit amounted to around 1700.[8] Many of these projects have been adjusted since their inception, scaling up the scope

[7] de Selding (2015b).

[8] According to the Union of Concerned Scientists Satellite Database. See Union of Concerned Scientists (2017).

of their services for most companies. A recent report published by Euroconsult further highlights the role played by these constellation projects in the future dynamics of the smallsat launch sector.[9] It is projected that 6200 smallsats could be launched over the next decade, and the deployment of these constellations is expected to be a driving force behind this forecasted demand.

The upcoming booming commercial market is expected to concern principally remote sensing and telecommunication constellations. These two segments will constitute the lion's share of the small satellites to be launched in the next years and are composed of a variety of purely commercial companies. On the other hand, historically, a consistent number of small satellites single missions have been represented by scientific spacecraft, technology demonstrator platforms both for civilian and military technology testing, as well as a number of military small spacecraft. This segment will continue and indeed *increase its share* but in more modest terms with respect to telecom and Earth observation.

As there is broad consensus among all major space stakeholders and firms that in the upcoming years the utilisation of small satellites will increase massively, correspondingly there is agreement that the business and revenues generated by these new systems will grow accordingly.

- According to Euroconsult, "*The total market value of these smallsats could reach $30.1 billion in the next ten years, up from $8.9 billion over the previous decade*".[10]
- MarketsandMarkets expects the small satellite market "*to grow from $2.92bn in 2017 to $7.53bn by 2022*".[11]
- SpaceWorks Enterprises' projections for the full market potential are 2600 nano- and microsatellites from 2017 to 2023.[12]
- The Business Intelligence and Strategy Research's report "Global Nano Satellite Market, Analysis and Forecast, 2017–2022" foresees a $6.35bn market for small satellites in 2021.[13]

At this stage, it is worth pointing out that forecasted demand and actual demand often differ due to a multitude of external environment factors influencing the launch business and its specificities, on top of the uncertainties about whether all these projected ventures will effectively achieve success or not. According to interviewed experts, the most important variables that are to be taken into consideration to predict how many smallsats will actually be launched are

(1) The robustness of the deployment strategies of constellation projects;
(2) The degree of penetration of CubeSats and nanosatellites;
(3) The ability to better address smallsats' weaknesses;
(4) The availability of access to space services.[14]

[9]Euroconsult (2017).

[10]Including satellite manufacturing, ground segment and launch services. Source: Ibid.

[11]Markets and Markets (2017).

[12]SpaceWorks (2017).

[13]BIS Research (2017).

[14]As outlined, inter alia, by ESA studies in 2016.

The intertwined relations between the booming digital economy and the small satellite industry will certainly have an impact on the future prospects of the latter, and this unprecedented access to real-time space data has implications that are still difficult to fathom. It is unclear as of yet whether EO will lead to significant growth in downstream applications, but the number of stakeholders involved in this line of business keeps on growing. Not only the ongoing developments in the Silicon Valley such as the growing involvement of big data companies such as Microsoft, Google or Amazon that have access to substantial funding and now have a stake in the space economy, but also the evolution of financing mechanisms for space ventures shows how the envelope is being pushed in space innovation.

3.3 SmallSat Market Segmentation

The small satellite market appears to be extremely heterogeneous, owing to the broad range of satellite classes spanning from dozens of grams to hundreds of kgs; an increasing number of actors, both public and private, that today are able to develop, build and launch a small satellite; and ultimately the great assortment of missions that smallsats can accomplish.

3.3.1 Mission Typology

When describing the market typology for smallsats, a first, obvious, a criterion of market segmentation is launch mass. This market is comprised of very different mass categories with payloads in the upper-end segment 500 times heavier than payloads in the lowest mass category. Although this can seem quite obvious given the mass range adopted for the category (up to 500 kg), it is interesting to point out that, with masses comprised between 500 and 20,000 kg, larger spacecraft present a more homogeneous group from a mass perspective with the heaviest payloads weighing a maximum of 40 times the lightest ones. Yet, in general, each launcher version is designed and optimised to be positioned on a rather limited market segment comprised of different spacecraft and mission profiles but within a narrow mass spectrum according to the orbit to be reached. Although various launch configurations exist with single or combined payloads, mass is an essential criteria to delineate target market segments of micro launcher projects and businesses. To provide for more homogeneous market segments, the ratio between lower- and upper-end mass was limited to 50 and the overall small satellite market divided into two distinct segments according to launch mass:

- **The nanosatellites market segment**, requiring solutions adapted to spacecraft with a mass of less than 10 kg (including in particular CubeSats, up to 6U).

- **The mini-/microsatellites market segment**, requiring solutions adapted to spacecraft with a mass between 10 and 500 kg.

This classification reflects the fact that the majority of smallsat spacecraft that have been launched since the exponential increase in the overall smallsat missions since 2013 belong to the class of nanosatellites (see Chart 3.3). However, it must be noted that this classification is a first-order approximation that serves the purpose of this analysis, but other classifications and mass bins could be established, based on mission types. For example, detailed mass analysis of smallsats projects reveals that the class below 50 kg is mostly populated by start-ups and university CubeSats (and notably, on top of these categories, also Planet's Doves). In the mass range between 50 and 150 kg, satellites are mostly employed for commercial purposes in radar, remote sensing and imagery missions, while in the 150–500 kg class lie the large-scale communication projects such as those from OneWeb and SpaceX.

Additional factors that can be taken into account to segment the small satellite launch market into more consistent markets include, among others:

- **Mission specifications**: Although 95% of small satellites are launched, or to be launched, to low Earth orbit, mission specifications can require an orbital injection in a range of different LEO orbits such as Sun-synchronous (SSO) and polar circular orbits and any other orbit fitting mission objectives. A distinction between missions with stringent orbital requirements and missions with more flexible orbital requirements can be made.
- **Constellations and single missions**: The small satellite market, and in particular the prospective market, includes a high number of satellites that are part of large constellations for remote sensing or broadband telecommunication. The launch strategy adopted to deploy, maintain and upgrade these constellations will differ from the strategy adopted for satellites that are part of single missions. The importance of constellations in the rationale for micro launcher initiatives also supports addressing them as separate customer segments.

Large Constellations: Remote Sensing and Communications

As noted, the majority of smallsats that will be launched in the coming years will constitute constellations spanning from dozens to thousands of spacecraft, most of them belonging to the commercial remote sensing sector, or even larger broadband telecommunication systems.

In the Earth observation and remote sensing sector, Euroconsult predicts the launch of up to 1100 satellites, and among them, four companies—Planet, DigitalGlobe, Spire, and BlackSky—constitute more than 85% of the total.

Table 3.2, extracted from the "*2017 State of the Satellite Industry Report*",[15] details several current and planned Earth observation and remote sensing commercial systems. The list includes systems whose satellites are already on orbit, or if funding, launch agreements or National Oceanic and Atmospheric Administration (NOAA)

[15] Bryce Space and Technology (2017).

licences have been announced. What clearly emerges from the list is that most of the current systems are constituted by few satellites of large mass, with the notable exception of Spire and Planet (somehow already belonging to the new trend), while the vast majority of upcoming new Earth observation (EO) systems will consist of a large number of small satellites ranging from a dozen to several dozen spacecraft.

The future landscape for commercial remote sensing missions in LEO looks crowded. In addition to the variety of existing remote sensing infrastructures, from the European Organisation for the Exploitation of Meteorological Satellites (EUMETSAT) and Copernicus systems to the US satellites and many others, new companies such as PlanetiQ, GeoMetWatch and GeoOptics are developing their commercial weather constellations, Planet is focusing on Earth observation imagery, and others such as Spire are looking also at offering AIS services in addition to imaging. Other examples are the BlackSky constellation from Spaceflight Industries (in partnership with Thales Alenia Space), which is to be composed of 60 microsatellites weighing 55 kg each and will start being launched from 2018 onwards targeting 2020 to complete their initial fleet, and the Finnish ICEYE start-up with plans for a 20 satellite constellation, that received $2.8 m Series A funding in 2015. The idea that the number of smallsat ventures competing on delivering the same kind of services is too high for the market gives more weight to the potential upcoming consolidation phase in EO and other sectors in the next two years. Indeed, since the beginning of 2017, a number of mergers in the EO market have taken place, following a re-assessment of many companies' business strategies: Planet acquired Terra Bella from Google, EagleView Technologies acquired OmniEarth and MacDonald, Dettwiler and Associates acquired DigitalGlobe to form Maxar Technologies. That so many new companies are trying to establish their presence and offer new services in the space remote sensing segment is undoubtedly enabled by the crucial combination of small satellite platforms' capabilities together with more options for space rides. This is further stimulated by the emerging big data economy enveloping and encompassing an ever-larger sector of traditional activities, for which space-based remote sensing data constitute a fundamental source.

Planet is an interesting case study to understand the dynamic behind deployment strategies of a remote sensing constellation. After the successful launch of 88 Doves in February 2017 with the Polar Satellite Launch Vehicle (PSLV) and 48 in July 2017 with a Soyuz vehicle, the latter being the last wave of initial deployment, Planet holds the record for operating the largest smallsat constellation composed of 190 "Doves". This gives the company the ability to monitor Earth's landmass on a daily basis.[16] In hedging against potential uncertainties inherent in the launch business, Planet has adopted a deployment approach that could be described as a "many eggs in many baskets" strategy, with regular launches, mainly as piggyback payloads until today. Working with a wider range of launch service providers and vehicles for launching their nanosatellites gives them a competitive edge on the market, as they are able to quickly respond to launch opportunities. Furthermore, the design of their nanosatellites allows streamlined production, and one Dove can be built within two weeks.

---

[16] Schingler (2017).

Table 3.2 Non-comprehensive list of current (green) and planned (blue/orange) commercial remote sensing constellations

Existing and Planned Remote Sensing Commercial Systems		
Company	**System Size**	**Sat Mass (Kg)**
Large satellites		
Airbus D&S	4	1000
DigitalGlobe	5	2800
MDA Radarsat	1	2300
UrtheCast	24	1400
Small satellites		
Spire Global	50	3
Terra Bella (Planet)	24	120
BlackBridge (Planet)	5	150
Planet	100+	3
DMCii	6	450
ImageSat	3	350
Astro Digital	30	20
Axelspace	50	95
BlackSky Global	60	50
GeoOptics	24	115
Hera Systems	48	24
ICEYE	50	100
PlanetiQ	12	22
Satellogic	300	35
Unknown Satellite Mass		
Capella Space	30	
XpressSar	4	
HawkEye 360	21+	
Planetary Resources	10	

Adapted from (Bryce Space and Technology, 2017)

Spire Global has adopted a similar strategy as they launch in small quantities, typically four to right CubeSats at the time, on a wide range of launch vehicles, mainly as piggyback or via cargo services to the ISS (Cygnus from Orbital ATK). Ideally, Spire indicated it would like to secure one launch every month. However, Spire's launch manager, Jenny Barna, has made it clear that the success of their business did not depend on micro launcher developments despite being willing to work with them.[17] Another example of smallsat constellation deployment is represented by a European smallsat company, ICEYE, which planned to launch its prototype satellites as a piggyback on PSLV and Falcon 9 launches and then announced the deployment of its entire constellation with Vector Space Systems' micro launcher. It is worth mentioning that these strategies, notably the exact number of required satellites, are evolving along with these companies' business models, as they also need to deal with the constraints of the launch industry.

With regard now to the satellite communications broadband market, according to Euroconsult 3100 small satellites are expected to be launched through 2026 with six major constellation projects, OneWeb and SpaceX being the most significant.[18] These new megaconstellations, intended to deliver worldwide broadband data coverage, are designed to be placed into LEO to ensure, among other advantages, a lower latency with respect to traditional GEO platforms. In order to achieve effective coverage and adequate revisit times, these constellations will require an unprecedented number of satellites, numbering in the hundreds or thousands; somehow, a feat made possible owing to the smallsat revolution. A non-comprehensive list of these projects is detailed in Table 3.3. However, it must be kept in mind that the commercial viability of these constellations is still to be determined.

The most advanced among these projects is that of OneWeb. Its constellation is composed of 720 satellites weighing 150 kg each, with plans to ultimately include 1280 additional satellites. The company aims at launching satellites from 2018 onwards to start commercialising their services on a global scale by 2020 and is considering expanding its initial deployment to further develop downstream applications in emerging digital sectors such as connected cars and Internet-of-Things (IoT). To deploy its constellation, OneWeb has signed a contract with Arianespace to launch 650–720 of their satellites on-board of 21 dedicated Soyuz launched from French Guyana and Russian Cosmodromes, with a launch frequency of one every five weeks.[19]

Supported by an investment from Google in early 2015, SpaceX is also poised to enter this LEO broadband arena, with plans for a 4250-satellite constellation. In November 2016, SpaceX filed an application with the FFC to launch its NGSO broadband Internet constellation and separately also proposed another 7500 second-generation satellites in very low Earth orbit to operate in the V-band.[20] The company launched prototype satellites in February 2018 and targets 2024 for full operational

[17]Foust (2016b).

[18]Euroconsult (2017).

[19]de Selding (2015c).

[20]Russell (2017).

Table 3.3 Non-comprehensive list of current (green) and planned (blue/orange) telecommunication constellations

Existing and Planned Commercial Satcom Systems in LEO/MEO		
Company	**System Size**	**Sat Mass (Kg)**
Large satellites		
O3b	12	700
Iridium	70	689
Globalstar	24	700
LeoSat	84-108	1250
Small satellites		
Orbcomm	18	22
OneWeb	720-2,720	Under 150
SpaceX	1,600-11,943	380
Unknown Satellite Mass		
Telesat	117-234	
Xinwei	32	
Yaliny	140	
Astrome Technologies	150	
KasKilo	288	
CAST	60	
Helios Wire	30	
Sky&Space Global	200	
Astrocast	64	
Kepler	140	
Lucky Star	156	
Viasat	24	
LaserLight	8-12	
O3bNetworks	48	

Adapted from Asia-Pacific Satellite Communications Council Newsletter (2017)

capacity. Worth noting is that SpaceX is planning on launching their satellites through its own Falcon vehicle.

Several more companies have presented their plans for their own constellations to the FCC. Boeing had planned on deploying two layers of constellations, of 2956 satellites and 60 satellites, but recently transferred its authorisation requests to OneWeb. Kepler Communications has plans for 140 satellites; Telesat Canada has filed for 234 satellites altogether, Theia Holdings A for 112 satellites, LeoSat MA for 80 satellites, and O3b constellations for 84 satellites. These are but the largest projects, and many other projects are also under development. Notably, in Europe, the Swiss Astrocast smallsat operator is targeting the IoT market with its 64 CubeSat constellation that recently raised \$3 m in seed funding from investors that focus on early stage technologies, including Airbus Ventures.

The commercial viability of such large-scale projects, so far existing at preliminary stages of development for most of them, remains to be seen. Even though the global economic and space industry landscape has much changed since the 90s—when the first constellation projects were launched, and, famously, ended up in bankruptcy—the ongoing evolution to a digital economy intersecting almost every aspect of human activities is creating an unprecedented demand for space data and connectivity.

When looking into the details of such constellations, it emerges that most new satcom systems envision hundreds if not thousands of satellites, particularly in the case of broadband constellations, in order to provide the appropriate coverage of the full Earth globe from a low orbit. Moreover, it is these systems are not comprised of nanosats or CubeSats, but of rather larger spacecraft, and their final orbital location will mostly need dedicated launches on board medium to heavy launchers for quick deployment, as traditional piggyback will certainly not satisfy mission requirements. Their deployment strategy will thus be significantly different from approaches adopted, for example, by remote sensing nanosatellite constellations.

Single Missions: Science, R&D and Replenishment

Before the exponential increase of smallsats used for commercial purposes and particularly in constellation configurations, which started in 2012, the market for small satellites was largely comprised of single missions, mostly scientific payloads and technology demonstrator missions (see Chart 3.3).[21] While the number of these missions will be dwarfed by the sheer number of upcoming remote sensing and telecom constellations missions according to the forecasts outlined above, the employment of smallsat platforms for science and R&D purposes will certainly continue to maintain—and increase, in relative terms—its relevance, owing to the lower costs, increasing popularity of this class of missions and growing opportunities to reach orbit.

Indeed, the importance of smallsat platforms for R&D and technology demonstrator missions for micro launchers is not to be understated. Before reaching the stage of operational space missions, whether scientific, military or commercial, basic technology to ensure reliable and high-performance solutions must be rapidly and affordably

[21] Wekerle, Bezerra Pessoa Filho, Eduardo Vergueiro Loures da Costa, and Trabasso (2017).

validated in the orbital environment. For this purpose, smallsat platforms constitute a major asset,[22] and indeed a great number of space industries and agencies employ smallsats platforms for R&D. One example among many is the NASA Pathfinder Technology Demonstrator (PTD), which is intended to test the operation of a variety of novel CubeSat technologies in LEO. The PTD mission is funded through NASA's Small Spacecraft Technology Programme (SSTP), which is charged with developing and maturing technologies to enhance and expand the capabilities of small spacecraft with a particular focus on communications, propulsion, pointing, power and autonomous operations.[23] In this sense, equally relevant is the employment of smallsats as R&D missions for the development of space-based military technologies and programmes. An example is the Space and Missile Defense Command-Operational Nanosatellite Effect (SMDC-ONE) spacecraft, launched in December 2010 as a secondary payload aboard a SpaceX Falcon 9 rocket with the Dragon capsule. The 3U CubeSat was the first of eight the US Army ever launched and helped demonstrate a beyond line-of-sight (BLOS) communications capability using small satellites in LEO.[24]

Scientific payloads on smallsats, launched either by public institutions or private companies, can encompass broad categories and purposes, collecting data for research ranging from astronomy, to material sciences, to biology. Indeed, on top of typical institutional research smallsats, a number of companies and start-ups have recently started exploring the possibility of using the unique microgravity environment of LEO for customised research applications by offering researchers CubeSat-sized laboratories. An example is SpacePharma, an Israeli biotechnology company, which has developed end-to-end miniaturized lab systems, provided with sensors and readers, capable of working on different microgravity platforms. These innovative platforms could be employed for research on pharmaceuticals, biotechnology, material and chemical science and nutrition.[25]

Interestingly, a number of scientific CubeSats will be embarked on the first flight of NASA's Space Launch System. The primary mission will send an Orion spacecraft to cis-lunar orbit, but after its release, 13 CubeSats will also be deployed from the rocket's stage adapter. These small spacecraft will thus benefit from the very rare opportunity of a moon-bound launch, an extremely rare event for piggybacked payloads. This rare event further underlines the—so far untapped—potential for deep space exploration by using small sats: one of them, NASA's Near-Earth Asteroid Scout (NEA Scout), is provided with a propulsion system and will conduct a two year and a half mission to explore a near-Earth asteroid.[26]

In addition to this range of single-satellite missions, as smallsat constellations are now being launched with ever more frequency, it can be expected that the replenishment of satellites in such constellations could become a relevant activity for launch

---

[22] See for example (Surrey Satellite Technology, 2017).

[23] National Aeronautics and Space Administration (2015b).

[24] Henry (2015).

[25] See SpacePharma website at http://space4p.com.

[26] Davis (2016).

service providers. However, in this regard, the class of satellites will make a huge difference in how smallsat operators will approach the objective of replenishment: constellations made up of nanosatellites (e.g. Planet's 3 kg Doves) and CubeSats, in particular, will definitely need a different replenishment strategy than mini- and microclass satellite constellations (such as OneWeb's 150 kg telecommunication smallsats). How smallsat operators will ultimately approach this topic remains an unknown factor, yet potentially having a large impact on micro launcher businesses.

3.3.2 Customer Typology

The characterisation of the small satellite market by mission typology outlined above must be complemented by a parallel characterisation distinguishing the customers of such missions. In this regard, three broad customer categories can be identified in the smallsat market: missions belonging to commercial satellite companies and operators; missions launched on behalf of government institutions, including defence; and lastly spacecraft developed by universities and research institutes as well as amateur projects.

Commercial Smallsat Operators

Commercial satellite operators, and here in particularly referring to smallsats, have been engaged for some years in the full variety of space missions outlined above. From deploying remote sensing and Earth observation to telecommunications, not leaving aside the all-important technology demonstrator and validator missions as well as for science, the booming smallsats market is undoubtedly led by commercial companies and commercial purposes.

Government and Defence

Other than commercial companies, governments and institutions have been a typical customer for smallsat missions, mostly for single-mission research and development purposes, as well as scientific payloads. In the defence sector, the most advanced space-faring nations have recognised the value of ever smaller satellites for military purposes, owing to their advantages in terms of stealth, increasing operational capabilities (e.g. reconnaissance and communication) and the lower cost of assembling and deploying a great number of smaller satellites into space.

In this regard, a November 2017 interview with the head of US Strategic Command, Gen. John Hyten, is highly relevant. In the interview, he acknowledged the rapidly increasing seriousness of threats to current US military spacecraft and argued for a necessary change of direction in how the US military procures satellites: rather than continuing relying on large, "exquisite" and costly systems that are essentially un-defendable, more resilient and distributed capabilities (i.e. smallsats) should be deployed. Examples of this trend are recent announcements from General Atomics about the development of defence-focused CubeSats,[27] and the agreement between the Finnish smallsat start-up ICEYE to supply airborne Earth observation data services to the US Defence Innovation Unit Experimental (DIUx).[28]

University, Education, Amateur Projects

To some extent, university and education smallsats could be considered a subcategory of government or institutional smallsat missions. However, for the purpose of characterising the smallsat market, and in particular assessing the various launch strategies and the value of micro launchers, it is appropriate to consider them as a separate category. The reason is that this type of customer typically does not possess the same financial resources as commercial satellite operators, or government customers, and thereby the price factor constitutes the main parameter to consider when addressing different launch solutions.

3.3.3 Characterising Customer Expectations

To complete the review of market forces and key trends identified above, a more detailed examination of the expectations of customers is required to understand how micro launcher solutions can provide viable solutions. The objective of this analysis of the demand is to understand, from a demand and supply perspective, the factors that drive the selection of a launch solution. The identification and characterisation of these factors, or criteria, will enable benchmarking different launch strategies vis-à-vis groups of customers sharing common launch service expectations.

Identifying a limited set of criteria driving the selection of a launch solution is not straightforward. Launching a satellite is a highly complex operation for which a high number of technical parameters and service features must be taken into account to assess the fit between the solution offered by launch service providers and customers' needs and requirements. This exercise is made more difficult by the variety of micro launcher projects and small satellite missions, which does not allow a thorough analysis of each specific criterion. Research and interviews with stakeholders, including micro launcher businesses and small satellite operators, have converged in identifying a high-level list of criteria to be taken into account to evaluate the fit between launch service provider's value proposition and customer's expectations.

On the small satellite launch market, the main factors influencing the selection of a launch service are:

- **Slot Availability**: time to the next available launch compatible with mission specifications. This criterion correlates the frequency of launch opportunities compatible with mission specifications and the customer's deployment strategy.

---

[27] See for example (Henry, 2017).

[28] Werner (2017).

- **Schedule Reliability**: ability of the launch service provider to guarantee that the satellite will successfully reach the target orbit on the fixed date. This includes a variety of factors, including calculation of the risk of manifest delays due to the domino effect after a failure.
- **Price-effectiveness**: the value-for-money for a smallsat operator to use a certain launch strategy.
- **Flexibility**: ability of the launch service provider to accommodate the customer's mission requirements with particular regard to orbital parameters as required by the mission type.

3.4 Competitive Analysis: Launch Solutions for Small Satellites

The diversity of launch options for smallsats has been increasing in recent years, responding to the emergence of a growing smallsat market.[29]

In fact, if one looks at traditional launch vehicle capabilities and capacities, they usually exceed the requirements of the primary customer. In particular, in almost any space launch, there is additional residual mass, volume and, in some instances, other performance margins to further accommodate one or more additional small spacecraft additional to the main payload. To exploit this excess capacity, a variety of adapters, dispensers and deployment technologies have been developed to house multiple spacecraft on top of the primary payloads of a given launcher. This approach, typically called piggyback, is currently the primary strategy employed to launch small satellites into orbit at a cheap price. Indeed, launcher companies and launch services providers are developing evolved systems and methods to meet the growing demand for access to space from the growing smallsat market.

Another opportunistic strategy is deployment from the International Space Station. The completion of the ISS over the past decade, and the addition of specific modules to its configuration, has created the opportunity to deploy small satellites—specifically, CubeSats—from the ISS itself through specialised companies like NanoRacks, after their delivery to the Station by means of one of the many resupply missions that are routinely launched into orbit to the ISS. Even more recently, some companies have been exploiting the possibility of deploying small satellites also during the orbital descent of such capsules, after docking with the ISS.

Both these approaches have resulted in a relatively cheap (when not simply free) orbital ride for small satellites, opening up new business opportunities for a variety of new smallsat companies for which the high cost of accessing space constituted the main barrier to developing their businesses. However, as smallsats become more and more capable, and these smallsat companies greatly expand their ambitions and business plans, the inherent limitations of rideshares and ISS deployment is becoming evident.

---

[29]Foust (2015).

For this reason, a new approach to orbital delivery that is gaining traction is a rideshare launch. This strategy refers to the launch of multiple satellites of which none is predominant, mutually agreeing on a compromise for the orbit/launch date. In certain instances, a company acting much as a broker between the launch service provider and the smallsat companies books a complete launch and sells the available capacity to multiple operators, without identifying a primary customer.

And lastly, as small payloads continue to expand their capabilities for a variety of applications and missions with specific requirements (orbital parameters, one-off quick deployment, interplanetary trajectories, precise rendezvous times and so on), deploying these spacecraft as primary payloads of a dedicated launch might indeed result in being the best, if not sole, strategy.[30] It is mostly to cover these emerging requirements and to seize this specific market opportunity that the new micro launcher initiatives worldwide are being developed.

The following chapters will discuss in detail the advantages and disadvantages of each approach from the perspective of smallsat operators.[31]

3.4.1 Piggyback

Piggyback is defined as the launch of one or more small satellites as secondary payloads taking advantage of excess capacity on a certain rocket, alongside a primary payload that is the main customer of the launch.

This method allows smallsat operators to have their hardware launched for very competitive prices, if not even for free in certain instances, provided they fulfil all the safety and physical criteria for being hosted on the launcher, and granted that the intended orbital destination is compatible with their mission requirements.

However, the main drawback of this approach is that piggybacked customers have little or no influence over essential launch and orbital parameters (primarily schedule and trajectory, these being unavoidably selected by the primary customer of the launch). In addition, currently, they often have to wait for long periods until excess capacity manifests on an upcoming launch compatible to their needs.

Yet, the discount prices offered by launch service providers to piggybacked satellites can be extremely appealing for a variety of smallsat developers with limited funding, such as amateur-led projects and university CubeSats, but not limited to these. Indeed, a growing number of commercial smallsat companies have started to massively employ these opportunity rides as an integral part of their satellite deployment strategies, benefitting from the reduced prices of accessing orbit and the overall reliability of the current fleet of launch vehicles. Furthermore, as smallsats—particularly microsatellites and below—are cheap and fast to produce, when deploying large constellations of those satellites, a loss of a certain number of payloads, while

[30]NASA Ames Research Center (2015).

[31]Niche deployment methods, such as deployment of a small payload from a “mother” satellite, are not discussed in this report.

certainly impacting the business of the company, does not necessarily result in a crippling blow for the business.

In fact, together with the increased use of smallsats and the reality that no dedicated micro launcher initiative has yet reached maturity, in recent years, the offer for rideshare opportunities has grown steadily.[32] An increasing number of well-established service launch providers have shown willingness to capture market shares in this segment, which in turn present smallsat developers with an increasing number of possible rideshare launch solutions, somehow mitigating the main disadvantage of having suboptimal schedule planning.

Almost all small-to-large-sized launcher vehicles (with a payload mass of more than 500 kg to LEO, as defined previously) offer today the possibility of accommodating secondary or small payloads, most often through dedicated dispenser and other auxiliary platforms.

The Indian PSLV is a prime example of how piggybacking is increasingly gaining importance as a solution for delivering small payloads to orbit. In February 2017, it set a new record in the number of simultaneously launched satellites by launching 103 small satellites aside a main payload, and another 34 in June 2017. As the Indian Space Research Organisation (ISRO) is planning to privatise PSLV launch services to increase its launch rate,[33] the Indian option will continue to be a major vehicle offering affordable rideshare opportunities over the coming years. Furthermore, consideration is currently being given to a new, smaller launch vehicle reportedly costing one-tenth of the manufacturing costs of a normal PSLV, with a total payload capacity of just above 500 kg to 500–700 km in altitude (SSO or LEO).[34]

Europe has demonstrated its willingness to enter this specific market segment with the first signed contract for Vega C for a launch planned in 2019,[35] as part of the Small Satellites Mission Service (SSMS) initiated by ESA in 2014.[36] This new capability, which will enter into service in 2019, will allow the highly reliable and versatile Vega rocket to provide low-cost and regular launch services for European Institutional light satellites and some commercial customers.[37] The SSMS dispenser with its modular design enables Vega to provide launch opportunities for light satellites with an overall mass ranging from 1 to 400 kg smallsats, with different alternative configurations and relevant combinations under a "rideshare" concept.[38] The re-ignition capability of the VEGA last stage (Attitude Vernier Upper Module—AVUM) enables smallsats to be separated on up to two different altitudes in the inclination of the SSO. A rideshare service on Ariane 6, called Microsat Launch Share (MLS), is also under development with ESA to take advantage of spare capacity on regular Ariane 62 launches to LEO by embarking multiple

[32]Foust (2017c).

[33]Government of India—Department of Space (2016).

[34]Singh (2017).

[35]ArianeGroup (2017).

[36]European Space Agency (2017b).

[37]Ibid.

[38]Ibid.

microsatellites at a very low price several times a year starting from 2021. Both European initiatives are being coordinated under the Low-Cost Lightsat Launch (LLL) initiative approved at the ESA Ministerial Conference in 2016, in order to foster standardisation and reinforce Arianespace's offering to microsatellite customers with regular launch slots on the renewed European launcher family.

3.4.2 Deployment from the ISS and Orbital Capsules

Another development that enables smallsats—and in particular CubeSats—to reach their orbital destination is to be released directly from the ISS. These satellites are first transported to the station via a standard launch of one of the several resupply cargo missions (such as Cygnus, Dragon, HTV) that periodically dock with the ISS and from there manually released into LEO through an airlock system operated by astronauts.

This solution has become increasingly popular over recent years, and it is for the most part operated by the company NanoRacks, which in addition to a variety of services offered within the space station itself, also offers the possibility to release payloads into orbit (see Table 3.4). The company offers to deploy CubeSats from their NanoRacks CubeSat Deployer (NRCSD) located in the Kibo module of the station,[39] microsats from the NanoRacks Kaber Microsat Deployer (Kaber), and has recently started offering CubeSat deployment from an external dispenser module located on the Cygnus spacecraft itself.

In the first solution, the NRCSD is capable of holding six CubeSat Units—allowing it to launch 1U, 2U, 3U, 4U, 5U and 6U (2×3 and 1×6 configurations) CubeSats. In contrast, the Kaber module can accommodate the 24U form factor, with a maximum payload mass of 82 kg. Lastly, CubeSats can be deployed from the Cygnus cargo capsule after it departs ISS, accommodating a 36U total volume and 1U–6U linear form factor, to a 51.6° inclination, 500 km (above the station). According to the company, owing to a higher orbit with less atmospheric drag, this solution can add approximatively two years lifetime for typical CubeSats, if compared to deployment from the Kibo module.

Releasing CubeSats from the ISS has become a streamlined and established process, now routinely part of the bigger picture of ISS operations. In particular, the company works closely with the NASA Payload Office, and satellites to be deployed through NanoRacks are reviewed by NASA's Payload Safety Data Panel. Furthermore, NanoRacks released standards for their customers to build their satellites and in so doing reduces the chances of payload integration issues. The CubeSat format is particularly fitting for the design of the spacecraft carrying them. NanoRacks offers an array of prices, starting from \$15k for educational projects, \$85k for a commercial 1U CubeSat, and higher for larger payloads.[40] The affordability of the service allows

[39]eoPortal Directory (2016).

[40]For more information see NanoRacks FAQ at: http://nanoracks.com/resources/faq/.

Table 3.4 CubeSats deployed from the ISS by NanoRacks. *Source* ESPI

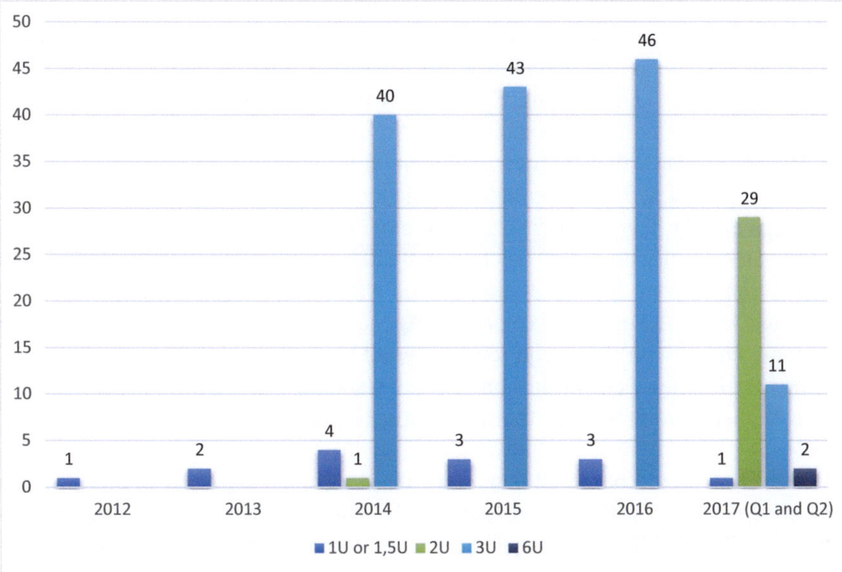

also cheap technology demonstrations for smallsat operators, as was the case, for example, for Spire Global.[41]

Nevertheless, this deployment option is constrained mainly by two factors: the limited capacity of the Kibo's airlock (which constrains the satellite designs) and the number and frequency of resupply missions that can host CubeSats for deployment. As a result, the number of satellites per mission is rather limited, potentially forcing customers to wait up to 14 months. In addition to that waiting time, it may take another 1–3 months after berthing before release.[42]

Additionally, perhaps the biggest disadvantage of this strategy is that the satellites are released in an ISS-type orbit (51.6°, approximatively 400 km altitude), which is a less than optimal orbit for many applications, such as remote sensing. This drawback is even more relevant when considering that smaller CubeSats do not typically have their own propulsion systems to move to different orbits.

However, the high popularity of this deployment strategy for smallsats suggests that it will continue to have a relevant role for newcomer smallsat developers, in particular in the light of the progressive transition of the ISS to more commercial utilisation.

In this regard, NanoRacks is planning to scale up its business in the medium term and expand its CubeSat deployment offer. While it has optimised the use of the ISS airlock, the company is still restricted by the size of the hatch in use. In

[41] Foust (2016b).

[42] Martin (2016a).

2016, it signed a Space Act Agreement with NASA to develop the first private airlock module on-board the ISS. The airlock will be unmanned and will be able to be removed from the station to be used on a future full-fledged commercial space station. Understudy is the potential reuse of second stage modules to build such new infrastructures. The Boeing Company is in charge of building the Airlock's Passive Common Berthing Mechanism used to connect modules on-board the ISS. NanoRacks estimates that the new commercial airlock will be launched in 2019, effectively increasing their offer in this segment of smallsat launch market. This new installation enables them to increase the number of airlock cycles they can perform and ultimately deal with larger batches of smallsats and/or bigger satellites. Worth noting is that the long history of cooperation with national space agencies such as NASA and Japan Aerospace Exploration Agency (JAXA) enables the company to align more easily with their requirements—and vice versa, the company has been pushing to reduce legal constraints on smallsat launches—and to have a more streamlined administrative procedure that facilitates their business and reduces overhead costs.

However, it should be noted that, since these resupply missions to the ISS take place under specific NASA procurement, delivering CubeSats to orbit through this method entails an additional layer of complexity for the operators, in terms of authorisations and processes.

3.4.3 Rideshare

A rideshare launch of small satellites is an emerging launch strategy involving the delivery of large batches of satellites owned by one or more customers employing the full capacity of one existing small or medium launcher, in some instances without a single payload identified as major customer.

The evident advantage of this strategy is the greater flexibility and schedulability offered with respect to a piggyback solution, as multiple customers of a single rideshare can mutually agree on orbital launch parameters and schedules.

Moreover, from the perspective of a smallsat megaconstellation operator, this approach could be particularly appealing to quickly deploy its large system, as in this way it will be able to do so with optimal schedule and mission control without having to depend on opportunity rides, with all the typical issues outlined previously. This approach is becoming increasingly relevant since, for those companies that are about to enter the smallsat large constellation segment, timing is of prime importance for delivering their services and thus generating revenue.

Increased availability of launch services has also given rise to new types of businesses including secondary payload integrators such as Tyvak, TriSept and launch services brokers such as Spaceflight Industries.[43] The latter in particular is one of the first companies to offer this kind of "launch service brokerage" approach. The

[43]Barna (2015).

company was considering performing its first launch of this kind in late 2017 with a payload adapter called Sherpa attached to a Falcon 9, but ultimately had to cancel due to the slipping schedule of the launch service provider.[44]

In Russia, the company Glavkosmos planned to launch 120 smallsats on a Soyuz 2.1v in 2017,[45] with the objective of further increasing availability for such opportunities. Their solution offers a certain degree of flexibility as demonstrated in July 2017, with the launch of 73 satellites into three different orbits.[46] This rideshare launch opportunity provided by Soyuz gives an interesting overview of smallsat trends as several constellation developers shared the same launcher, such as for instance, Planet, Spire Global, GeoOptics and Astro Digital.

As launch vehicles develop ever more refined capabilities to launch multiple smallsats through dedicated adapters and dispensers, this approach will constitute an important new strategy for orbital smallsat delivery in the future.[47]

3.4.4 Dedicated Launch in Micro Launchers

As the capabilities, numbers and market prospects of small satellite platforms increase significantly, the expectations and needs of small satellite companies in terms of accessing space also become more sophisticated. As a consequence, current approaches such as piggyback or rideshare and ISS deployment, while economically affordable and presenting increasing availability for small satellites, in certain instances might result in less than optimal approaches for the growing needs of operators.

It is in this context, and to address these increasing requirements, that most commercial micro launcher initiatives are being developed. In offering a potential schedule comprised of numerous launches per year dedicated exclusively to smallsats, as well as total control of the mission parameters, micro launcher companies aim not only at filling the gaps in current access to space solutions, but also at opening up new possibilities for space activities to be exploited by smallsat developers.

However, similar to the non-homogeneous demand side of small satellites as highlighted in Sect. 3.3, the micro launcher supply side also exhibits a great variety of payload performances to orbit, ranging from a few kgs to hundreds (see the projects overview in Sect. 2.4.1). This is important to highlight, as different vehicles will be thus able to address different segments of the market. Notably, most advanced launcher commercial projects lie in the range of 150–500 kg to LEO and would consequently be suited either to launch a single small satellite or even multiple mini- or microsatellites.

[44]Foust (2017d).

[45]Foust (2017b).

[46]Richardson (2017).

[47]Pultarova, Big launch companies predict doom for upcoming smallsat launchers, (2017b).

For this reason, it is relevant to stress that micro launchers, despite their limited capacity, could also be suited for deployment of several satellites at once. This is the case of Rocket Lab's Electron, which is scheduled to launch 20–25 of Planet's Flock nanosatellites in several launches, starting in 2018.[48]

3.5 Benchmarking on Smallsat Launch Options

As discussed in Sect. 3.3.3, the main factors influencing any small satellite company and developer looking for launch opportunities are availability and schedule reliability of the launch, control over the mission requirements and orbital parameters to properly fulfil the mission objectives and, last but not least, the all-encompassing factor of economic affordability.

Each factor detailed in Sect. 3.3.3 is qualitatively assessed with regard to the offer and performance of the different orbital launch strategies discussed in Sect. 3.3 for a typical smallsat mission profile, in a relative scale of three levels from "very low" to "very high". The assessment is presented in Table 3.5. The assessment of these parameters is based upon a range of interviews conducted with stakeholders, as well as comparative analysis of literature and historical data, launch service manuals, etc. It is important to highlight that both rideshare and launch on micro launchers are emerging approaches, therefore this assessment is based on the available information through companies' announcements, business plans and forecast sector evolution.

When considering **Availability**, solutions in which small satellites are to be accommodated as secondary payloads on an opportunity basis retain a medium score, owing to the limited number of global yearly launches; and due to the fact that not all these launches can accommodate all kinds of small satellites, nor can their mission profile fit the mission parameters for all classes of smallsat missions. Rideshare solutions achieve a high score, as they possibly require booking the entire capacity of new small-to-medium rockets, even though such solutions are still uncommon as of today. Additionally, as a matter of comparison for dedicated micro launchers, while the typical launch procedure of a small-medium sized rocket could take several weeks to months, the air-launched Pegasus XL could be launched in as little as a week's notice.[49]

The assessment of **Schedule** is essentially based on the risk of encountering delays (or failure) in the launch of the small satellite(s). This parameter is considered low for rideshares inasmuch as it is common that the launch manifest of established launch service provider's experiences reshuffles or delays. Vice versa, both micro launchers and rideshare promise to have high reliability in terms of schedule. However, regarding micro launchers, it must be recalled that they are experiencing years of delays during their development phase and given the extremely ambitious

[48]Krebs (2017).

[49]Taverney (2017).

Table 3.5 Assessment of current and emerging space launch strategies for smallsats with respect to customer expectations

Benchmarking Launch Strategies for Smallsats				
	Piggyback	Orbital Capsules Dispensers and Deployment from ISS	Rideshare	Dedicated Launch on a Micro Launcher
Availability	Medium	Very low	High	High
Schedule	Low	Medium	High	Very high
Price-effectiveness	Very high	Medium	Medium to high	Medium to low
Flexibility	Low	Very Low	High	Very high

launch rate increase hypothesis they have been generating, it is possible they will encounter delays for their prospective customers.

Price-effectiveness as a parameter represents the value-for-money for a smallsat operator to use a certain launch strategy. The assessment considers opportunity rides with high cost-effectiveness, owing to the cheap prices allocated for small satellites filling unallocated space along primary payloads (sometimes even offered for free). On the other hand, ISS deployment solutions retain a "medium" assessment based on the different pricing policy of a provider such as NanoRacks, distinguishing between commercial or educational payloads (see Sect. 3.4.2). Lastly, rideshare's price-effectiveness is expected to be high, when considering typical price per kg of small-to-medium rockets, whereas micro launchers do have a tentative medium to low price-effectiveness, in the light of the expected launch service price and considerations related to the filling rate of the rocket (see Sect. 2.4.2).

Flexibility on the selection of mission and orbital parameters is naturally low for those launch solutions in which the smallsat operator has little to no choice, as they are either decided by the primary payload of a piggyback ride, or locked by the orbit of the ISS. By contrast, rideshares and, even more so, dedicated micro launchers offer the highest flexibility in this regard.

However, it must be noted and specified that the selection of a specific launch solution over another will ultimately depend on the relative weight different customers assign to the various benchmarked criteria. Clearly, the resources and priori-

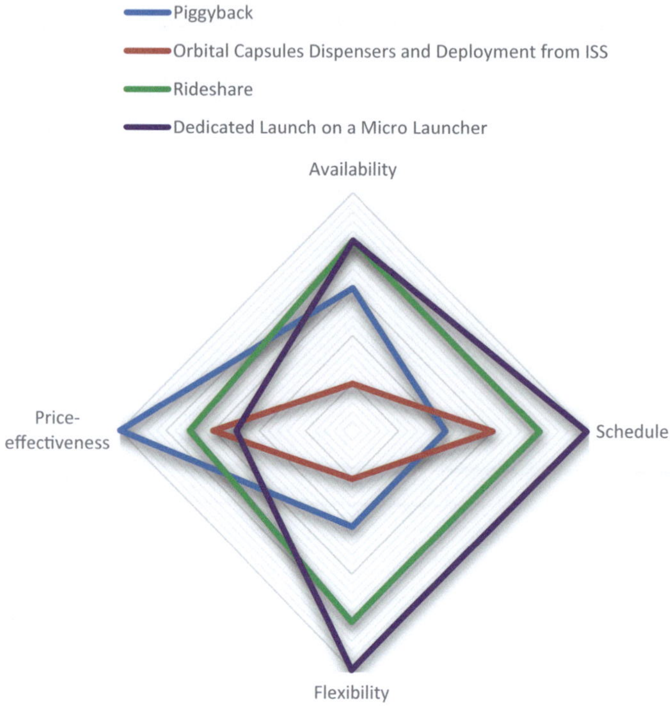

Chart 3.8 Benchmarking on launch strategies for smallsats according to the four criteria outlined in Table 3.5

ties of a university CubeSat project are not the same as those of a megaconstellation operator.[50] More specifically, a commercial company might put a higher priority on rapidity of deployment for its constellation, in order to generate revenue as quickly as possible, while keeping the price factor under reasonable limits; scientific research satellites might have less pressing needs for rapidity of deployment but more specific requirements for orbital injection; universities will usually look for the cheapest solution and rank a low price before any other launch service provision; and for military missions, a greater focus will be put on fulfilment of all mission objectives including reaching the target orbit in the timeliest fashion no matter the price of launch (Chart 3.8).

[50]Indeed, this is reflected in NanoRacks pricing policy, which starts from $15k for educational projects, and $85k for commercial 1U CubeSats.

3.6 Value Proposition of, and Stakes for, Micro Launchers

The value proposition offered by micro launchers as discussed in the comparative assessment in Sect. 3.5 are, in principle, clear. By proposing a launch perfectly suited to the customer, constituted by a potential on-demand schedule comprised of numerous launches per year dedicated exclusively to smallsats, as well as total control of the mission parameters, the companies aim not only to fill the gaps in current access to space solutions, but also to open up new possibilities for space activities to be exploited by smallsat developers.

The range of advantages of micro launchers described above—yet to be proven, as no micro launcher has reached the stage of full commercial exploitation as of February 2018—does naturally come at a price. Historically, price per kg to orbit is generally inversely proportional to the capacity of the launcher itself (see also Sect. 2.4.2). Although developed at a different time, the very first launcher to offer a micro launches-sized capacity to LEO was the air-launched Orbital ATK Pegasus: as described in Sect. 2.1, it achieved relative success, yet when compared to other launch options, it was also one of the most expensive vehicles in terms of price per kg to orbit. The technological landscape for rocket manufacturing has certainly evolved over the past thirty years, and a variety of new technologies, processes and engineering approaches have the potential to reduce the price per kg to orbit for a micro launcher.

Furthermore, as small satellites are relatively cheap and easy to manufacture, with fewer constraints on failure rates, their cost is noticeably lower. This raises another question for micro launchers: while for larger rockets the value of the primary satellite typically is much greater than the price of the launch; for micro launchers, the value of the small satellite(s) on board, not considering insurance where applicable, could well be on par if not less expensive than the launch itself.

Extremely cost-effective production, integration and testing of the launch vehicle, its subsystems, systems and the system integration are all mandatory factors for micro launchers to reduce the cost of the launch (in absolute terms).

As shown in Sect. 2.4.2 and Table 2.2, most micro launcher companies advertise prices between \$20k and \$40k per kg, noticeably lower than the "previous generation" of micro launchers (such as Pegasus XL and Minotaur-1). However, it must first be noted that, as mentioned above, these prices are the most optimistic baseline where the full capacity of the rocket is employed, thus ignoring fairing optimisation.[51] Even more important, to achieve those prices, a very high launch frequency is planned for the vast majority of them (see Chart 2.1), in order to generate economies of scale that can recover and amortise the development costs.

Indeed, keeping the price of launches low, together with the effective materialisation of the expected smallsat booming market, and the structure of competition from current small and medium launchers, will be key in determining the success of micro launcher initiatives.

---

[51] While this is true also for the reported prices of other rockets throughout this report, fairing optimisation might result in being less of a problem for larger rockets.

This consideration leads to the question of how precisely the still-to-enter-into-service micro launchers will intercept the still-to-be-fully-developed commercial smallsat market or, more specifically, which smallsats micro launchers will effectively launch to cope with their expected high launch frequency.

Indeed, with their rather limited payload capacity, micro launchers will not be the most efficient way to deploy the largest upcoming commercial smallsat large constellations. In particular, this applies to the telecom broadband constellations that, according to initial plans, comprise many hundreds of mini and microsatellites across multiple orbital planes.

For these missions, a rideshare approach (by employing the entirety of medium to heavy sized rockets) will most certainly be more effective in the deployment of constellations, as their initial focus will be on a timely deployment of the whole system to quickly generate business and revenue. Subsequently, demand will veer towards constellation replenishment, with a focus on reaching the specific orbits of the system. In this segment, micro launchers could play an important role. However, this option would be weighted by operators with respect to the possibility of simply launching spare satellites (as they are relatively cheap to produce), or replacing complete orbital planes, two solutions for which a rideshare would be preferred.

With both micro launchers and—even more so—megaconstellations still to materialise, it is premature to quantitatively assess this factor. Nonetheless, dedicated replenishment for large constellations is one of the possible future market segments in which micro launchers could play a role.

Setting aside this aspect pertaining to very large smallsat constellations, the added value of micro launchers lies in terms of their availability and flexibility with respect to traditional launch strategies. Such added value would in principle still strongly appeal to a variety of smallsat customers, particularly to those for whom the low cost-effectiveness factor does not represent a particular drawback when compared to the other parameters.

In fact, the high cost of accessing space is perhaps the main blocking factor for emerging smallsat commercial companies, yet timely access to space could also become a make-or-break element for those companies, particularly start-ups, to quickly start to generate revenue and thus survive in an increasingly competitive market for space-based services.[52] In this regard, micro launchers might be favoured by customers who cannot afford delays on their launch schedule (provided that they can reach their ambitious launch rate targets) and/or need to go into specific orbits not usually serviced by traditional launches in order to provide their services and fulfil their mission.

[52]Indeed, time-to-orbit is of high relevance for those smallsat start-ups who have raised funds from investors and thereby are compelled to maintain the forecast they have presented them.

This, indeed, could well be the case for defence-oriented spacecraft, in addition to commercial ones.

It must be noted that, as for every spacecraft involving military-grade technology, in this case the launcher would necessarily have to undergo specific certifications and fulfil clear requirements, the like of which would not be necessary when addressing the purely commercial market. However, for their ability to achieve quick and responsive access to space, micro launchers clearly continue to attract a significant interest from the defence sector (see Sects. 2.3.2 and 3.3.2).

Regarding technology demonstrator spacecraft, unless the specific mission has to be performed in particular orbits to accomplish its objective, traditional launch options—from opportunity or rideshares—would ultimately prove more affordable, regardless of the customer of origin. A similar argument can be made for small-scale scientific missions and even more so for university and education projects.

Here, perhaps lies the conundrum for micro launchers: the customers who would benefit the most from the added value offered by micro launchers (and with pockets deep enough to stomach the typically higher price per kg to orbit) are those who do not need to launch as often as would be required to achieve the very high launch frequency anticipated to make their business cases—considerations of future elasticity of the market set aside.

To conclude, as has emerged from the research and extensive interviews conducted with a variety of stakeholders, while the market for micro launchers remains as of yet unclear, a few considerations can be identified.

Assuming that the micro launcher space transportation offer will exhibit a price structure akin to that outlined in Table 2.2 in terms of both absolute value and price per kilogram, and bearing in mind individual value propositions, and the current and evolving offers of established launch service providers and their small-to-large vehicles, it is safe to argue that:

- Commercial customers will opt for rideshare solutions for the deployment of large-scale constellations (in particular for remote sensing and satcom, i.e. in 150–500 kg class of satellites) and could possibly consider dedicated launch of a micro launcher vehicle for replenishment of smallsats in the mini-/microclass segment in the longer term.
- Institutional operational missions (in particular defence) could favour micro launchers for single spacecraft with very specific orbital and schedule requirements. When launching single payloads with less stringent mission requirements

(such as technology demonstrators, R&D, science), a rideshare or piggyback approach could offer the best value-for-money.

- CubeSats from universities, small-scale science, as well as amateur or educational projects, will continue to favour opportunity rideshare or deployment from the ISS despite long manifest waiting times and limited choice on mission parameters, as they will continue to represent the most affordable solution.

This reflection is summarised in Table 3.6.

Table 3.6 Added value of, and stakes for, micro launchers across small satellite market segments

Added value of and stakes for micro launchers Across small satellite market segments				
	Main customer requirements	Strengths of micro launchers	Weaknesses of micro launchers	Stakes for micro launchers companies
Commercial remote sensing and communication constellations	• Timely access to space • Specific orbits	• Availability • Schedule • Flexibility	• Limited payload capacity for large deployment • Price	• Constellation deployment more efficient through small/medium sized launchers • Competition with rideshare offers
Replenishment of smallsat constellations	• Specific orbits	• Availability • Flexibility		• Replenishment is a niche option and depends on the success of mega constellations
Science, R&D, technology demonstrator	• Low price	• Availability • Schedule • Flexibility	• Limited payload capacity for large deployment • Price	• Limited number of missions • Competition with rideshare and piggyback offers
University, education, amateur CubeSats	• Low price	• Availability	• Price	• Hardly a business case per se, but they add to the basket of available missions • Price typically too high for this class of customer
Institutional (defence)	• Ready and targeted access to space • Price not an issue	• Availability • Schedule • Flexibility		• Limited number of missions • Could be partially addressed by non-commercial vehicles • Necessity for launcher certification for military-grade payloads

Chapter 4
Policy Perspectives for Europe

This chapter aims to assess what the stakes for Europe are in the micro launcher sector. The chapter first assesses whether micro launchers are a necessity for Europe by elaborating on demand and supply conditions and, building on that assessment, subsequently provides some policy considerations for pan-European stakeholders.

4.1 Micro Launchers in Europe: Status, Drivers and Stakes

Continued and renewed interest in micro launchers in Europe is exemplified by the range of privately led initiatives aiming at the commercial smallsat market, on top of a smaller number of institutionally led projects more geared towards the objective of developing innovative launch technology demonstration and validation. The following paragraph presents a detailed overview of European initiatives as of beginning 2018.

4.1.1 Supply: Status, Outlook and Funding

Table 4.1 provides an extract from the global worldwide overview presented and discussed in Sect. 2.4.1, here limited to the European context. As in the analysis of worldwide developments, the vehicles are classified in two categories by their **reference market** (commercial or institutional), the visible **financial backing** of the projects and their **status** (indicating the estimated level of maturity of the initiatives as of early 2018).

As emerges from this overview, various projects have already been announced or initiated in Europe with several of business models and concepts, demonstrating a clear interest from private and public stakeholders in the trend (even more so when taking into account cancelled projects).

M. Tugnoli et al., *European Access to Space: Business and Policy Perspectives on Micro Launchers*, SpringerBriefs from the European Space Policy Institute,
https://doi.org/10.1007/978-3-319-78960-6_4

Table 4.1 Overview of European micro launcher projects

Overview of European micro launcher projects							
Country	Launcher name	Company name	LEO capacity (kg)	Reference market	Scheduled first flight	Visible financial backing	Status (early 2018)
France	Eole/Altair [a]	ONERA (and others)	50/150	Institutional		Public	Prototype tests
France	Daneo [a]	MT Aerospace/Dassault Aviation	75 (150 to SSO)	Institutional		Private	Studies
Italy	Primo	Leap Space		Institutional, Commercial		Private	Announcement
Italy	Quick Launch Vehicle	ELV		Institutional, Commercial		Private	Studies
Italy	Final Launch Vehicle	ELV		Institutional, Commercial		Private	Studies
Europe	SMILE		50–115	Institutional, Commercial		Public, private	Announcement
Europe	Sparrow	ArianeGroup	500	Institutional, Commercial		Private	Cancelled
Europe	Q@ts/Q@rk	ArianeGroup	50/10	Institutional		Private	Studies
Norway	North Star	Nammo	10	Institutional, Commercial	2020	Public, private	Ground tests
Portugal		Deimos					Studies
Romania			150				Studies
Spain	Arion 2	PLD Space	150	Commercial	2020	Public, private	Announcement

(continued)

Table 4.1 (continued)

Overview of European micro launcher projects

Country	Launcher name	Company name	LEO capacity (kg)	Reference market	Scheduled first flight	Visible financial backing	Status (early 2018)
Spain	Bloostar [a]	Zero2infinity	75	Commercial	2019	Private	Prototype tests
Spain	Sagittarius Space Arrow [a]	Celestia Aerospace	16	Commercial			Announcement
Sweden	Rainbow Smallsat Express	Swedish Space Corporation	150	Institutional, Commercial	2020	Public, private	Announcement
Switzerland	SOAR [a]	Swiss Space Systems	250	Commercial		Private	Cancelled
UK	Black Arrow 2	Horizon Space Technologies	350	Commercial			Announcement
UK	Orbital 500R [a]	Orbital Access	500	Commercial	2020	Public, private	Announcement
UK	Prometheus	SpaceLS	150–250	Commercial			Announcement
UK		Orbital Express	220	Commercial		Public, private	Ground tests
UK	Devon 2	Tranquillity Aerospace	4	Commercial			Announcement

Looking into detail, it is worth noting that:

- The UK is the country with the most micro launcher projects, as five companies are active in developing vehicles clearly targeting the commercial small and microsatellite market, funded by a mix of public and private capital. It is noteworthy that in 2014, the National Space Technology Strategy (NSTS) and Space Innovation and Growth Strategy (IGS) underlined the need for a low-cost domestic launch vehicle in the UK.[1] Closely related to this assertiveness in this field, it is also worth recalling (see Sect. 2.4.3) the on-going efforts to establish a spaceport on UK national territory.
- In several other countries, such as Norway, Sweden, Spain and Italy, one or more companies are developing vehicles also intended for the commercial market, all of them at studies level or very preliminary stages of development.
- In France, initiatives from ONERA together with CNES (and other European companies such as Nammo) are testbed projects for the development of new technologies dedicated to smallsat launch. It is worth mentioning that a number of early studies have taken place in the past decade in France with regard to the development of micro launcher vehicles, mostly targeting the institutional and defence market, which have however not reached an operational phase.
- Romania is actively involved in the development of micro launcher concepts and related technologies, owing to the country's long-time expertise and tradition in aerodynamics and areonautical research—mainly through its "National Institute for Aerospace Research Elie Carafoli" (INCAS) research establishment and the Romanian Space Agency (ROSA). In addition to being involved in the SMILE project, among other activities INCAS was recently awarded a Phase 0/A contract to study a micro launcher concept (with performance of 150 kg into a 600-km circular polar orbit, to be launched by a launching site located in Europe), within the framework of ESA's FLPP. The targeted market of this potentially competitive European-based vehicle is the microsatellites segment developed by universities, but also by small companies for research purposes as well as validation of applications/products before commercialisation.[2]
- SMILE, a pan-European project funded by Horizon 2020 grants and at early stages of development, aims at "*ensuring Europe's independent access to space for small satellites*",[3] further making use of the Andoya Space Center as a launch site.
- Among cancelled projects, the 500 kg to LEO Sparrow micro launcher was being studied by ArianeGroup.
- MT Aerospace (a subsidiary of OHB SE) is reportedly involved in several studies with other companies and entities, such as Dassault Aviation and Zero 2 Infinity.[4] More broadly, European companies have been looking for B2B and B2G synergies in the possible development of a European solution for micro launchers.

---

[1]UK National Space Technology Steering Group (2014).

[2]Afilipoae et all. (2018).

[3]For an overview of the SMILE project see the webpage: http://small-launcher.eu/project/.

[4]OHB SE (2018).

- All European projects are still at very early stage of development, which should not lead to an operational capacity before 2020, i.e. two years after the most advanced projects outside Europe (e.g. Rocket Labs' Electron).

As for funding, these European micro launcher initiatives are currently financed through a variety of mechanisms, provided by private investment capital, national government and European institutions.

With specific regard to public funding of European micro launcher initiatives, from a **national** perspective, some countries are allocating funding as part of domestic space strategies and programmes for the development of national micro launchers, often in relation to an expansion of existing launch infrastructures. This is the case of the Norwegian Space Center, which is partially backing the development of the Nammo's North Star Launch Vehicle.[5] Moreover, the UK Space Agency has awarded the British company Orbital Access a €275k grant to define a roadmap for future UK small payload launchers. Highly relevant in this regard is the new national industrial strategy of the UK, presented in late 2017, which includes around $67 m dedicated to the support of new launch sites and launch vehicles.[6]

From a pan-European perspective, both ESA and the European Commission have so far provided limited seed funding for micro launcher initiatives.

In the past, ESA has allocated funding to a number of companies, within its Future Launchers Preparatory Programme (FLPP), particularly with regard to propulsion systems technology (notably the North Star Launch Vehicle initiative, and the Romanian INCAS).[7] The Spanish company PLD Space has also received financial backing from the local government and ESA with a €750k grant within FLPP.[8] More recently, ESA has been supporting a number of studies investigating the feasibility of an economically viable, i.e. commercially self-sustainable, approach for access to space of satellites with a mass up to 200 kg. This includes identification of technologies that could be developed with possible co-funding in a future phase. In this regard, in February 2018 ESA announced that five European companies were awarded studies contracts to explore the potential feasibility of small launch vehicles as part of FLPP. The five companies are ArianeGroup, MT Aerospace, European Launch Vehicle (ELV), Deimos and PLD Space.[9] As noted by Jerome Breteau, manager of ESA's Future Launchers Preparatory Programme (FLPP) during this announcement, "*a European commercial micro launcher can meet the growing need for dedicated launch services to companies with small satellites*".[10]

The research grant scheme of the European Commission Horizon 2020 is also financing some micro launcher initiatives, in particular for the SMILE project as well as others, such as Altair, ONERA's Eole Project and Orbital Express, another

[5]Haemmerli, Boiron, and Verberne (2016).

[6]Foust (2017a).

[7]Haemmerli, Boiron, and Verberne (2016).

[8]PLD Space (2016).

[9]European Space Agency (2018).

[10]Ibid.

UK-based micro launcher company which was recently rewarded the Horizon 2020 SME Instrument Award in 2017.[11]

Some past and recent European micro launcher projects are described in greater detail the following paragraphs.

Sparrow, Q@TS

In a 2016 report by the French Government,[12] a concept for a micro launcher by ArianeGroup called Sparrow, studied under full private funding from ArianeGroup and an external European private investor, was briefly outlined. In the report, the project was described as a "*pilot project for the NewSpace in France and Europe*" and intended to "*promote space research through fast and cheap access to space*",[13] particularly in the microsatellite and CubeSat segment. Furthermore, it aimed at "*structuring the microsatellite/CubeSat market by defining of a standard in line with the launch offer*".[14] However, in late 2017 it was reported that the Sparrow project had been cancelled as its business case did not materialise as initially expected, and a decision to focus efforts on adapting Ariane 6 launch service offer to microsatellites including with a rideshare solution (MLS) was taken by ArianeGroup.

As part of the ESA-awarded studies for micro launcher concepts announced on 8 February 2018,[15] ArianeGroup is reportedly studying a nano launcher named Q@TS ("Quick Access to Space"), with a performance of 50 kg to LEO.

Vega Light/Quick Launch Vehicle/Final Launch Vehicle

As news about the cancellation of ArianeGroup's Sparrow started to emerge, the other major European company providing space transportation vehicles, Avio, announced early-stage studies for a micro launcher. In mid-2017, at the Paris Air Show, Avio CEO Giulio Ranzo announced that the company was involved in a preliminary R&D study of a scaled-down version of Vega, named Vega Light. In his words, "*the aim is to have a minilauncher for satellites weighing 300–350 kg. It could be a possible development to meet market demand*".[16]

In a more recent development, as part of the ESA-awarded studies for micro launcher concepts announced on 8 February 2018, ELV announced that two micro launchers vehicles are being studied, based on the Vega architecture, and reportedly called "Quick Launch Vehicle" and "Final Launch Vehicle".[17]

---

[11] Orbex (2017).

[12] Dedieu and Menetrier (2016).

[13] Ibid. p. 25.

[14] See footnote 12.

[15] European Space Agency (2018).

[16] ANSA (2017).

[17] European Space Agency (2018).

SMILE

The SMILE project (SMall Innovative Launcher for Europe) is another pan-European initiative to tackle the perceived need for small payload launch capacity in Europe. Thirteen public and private entities, including companies and research institutes in eight countries, are involved in a consortium to design the concept for a European launcher for small satellites that is currently funded within the framework of Horizon 2020 with a €4 m grant.[18] Stakeholders notably include the Netherlands Aerospace Centre (NLR), the German Aerospace Centre (DLR), Nammo Raufoss ASS and the Andoya Space Centre (Norway), Terma (Denmark), the National Institute for Aerospace Research or INCAS (Romania), Airborne Composites Automation (Netherlands), Heron Engineering (Greece), Innovative Solutions in Space (Netherlands), 3D Systems (Belgium), PLD Space (Spain), Tecnalia (Spain), BoesAdvies (Netherlands), and WEPA Technologies (Germany). In this sense, the SMILE project brings together several micro launcher developers, companies and space institutions across Europe.[19] The overarching objective is to develop a 50 kg to LEO payload capacity micro launcher concept that will be able to drastically drive down production and operating costs and meet the needs of the European Space Research and Technology Development (RTD) as well as potentially enter the commercial market for nanosats. The price tag targeted by the consortium is less than €50k per kilo, *translating to a launch price* of €2.5 m in total.

4.1.2 What Demand for European Micro Launchers?

As in other countries, the multiplication of micro launcher initiatives in Europe primarily builds on the substantial growth of the small satellite markets observed during the last five years and expected to be amplified in coming years by a number of trends including, among others, massive deployment of CubeSats and small sats megaconstellations, advancement in technology miniaturisation and a growing interest in small platforms from a variety of governmental and commercial actors.

When assessing the demand for European micro launchers, however, several outstanding issues inevitably come to the fore, casting some doubts on the actual solidity of the companies' business cases.

As discussed in the analysis of business perspectives (see Sect. 3), similar to their foreign competitors, European micro launchers face several uncertainties related both to the small satellite market and to competition from already existing solutions that may seriously limit the number of satellites and related mass that micro launcher businesses could actually launch.

For one thing, even though there appears to be wide consensus that the small satellite market will continue to dramatically expand in the near future (see Sect. 3.2),

---

[18] European Commission (2016).

[19] Oving et al. (2016).

it is still not possible to accurately forecast the size of the small satellite market in the next 10–15 years with a high degree of confidence.[20]

A second and more substantial issue when assessing the demand for micro launchers is the identification of what part of this growing small satellite market will be captured by micro launcher companies. It is unclear whether European micro launchers will be able to seize a considerable share of the small satellite market, mainly because will have to compete not only with micro launcher companies in other areas, but also, and perhaps more importantly, with the current solutions for launching small satellites, including piggyback solutions and rideshares.[21] Interestingly, this point was also raised by top executives of the leading launch service providers (Arianespace, SpaceX, ULA, and ILS) during the World Satellite Business Week in September 2017. When interviewed about the prospects for micro launchers, they more specifically commented that "*emerging companies developing a generation of dedicated small satellite launchers will struggle to compete in the market place and will ultimately lose out to the bigger players (as) the future satellite launch market will favour ridesharing and customised services on larger launch vehicles rather than tailored launches by the newcomers*".[22]

These considerations are closely related to another unsettled issue in the business case of micro launcher companies: the readiness of small satellite operators to pay the price of their launch services.[23] Although satellite operators have expressed a clear interest in micro launcher solutions given the limitations experienced with traditional launch solutions (i.e. payload restrictions, orbit restrictions, and schedule delays), there is still much uncertainty as to how the range of advantages offered by micro launcher companies will be weighed against the associated price level.[24]

Indeed, as emerges from the assessment performed in Sect. 3.5, the value proposition of micro launchers lies in their high level of flexibility, availability and schedule.

---

[20] As emerges from—among others—a recent ESA's analysis of small satellite market.

[21] See Sect. 3.4.1.

[22] Quoted from Pultarova (2017b).

[23] This might well be the case if some launch service providers undercut this limited market with free piggyback solutions.

[24] When interviewed about his take on micro launchers, Arianespace CFO Pierre Godart contended: "Obviously, there is an interest from customers in micro launchers but the question is whether there is a price, which would be compatible with what the customers expect but also feasible for the micro launcher operators. We would obviously do that if we believed that there is a market. For the time being, we are working on studies. There has been no final decision yet. It is a debate among the industrial players, whether the market is there. The operators are saying that they are interested but that's not enough. The question is what would be the launch price, what would be the cost of the development, what would be the potential margin that you can realize. There must be, at the end of the day, a business case. In our opinion, it is maybe too early. None of the big constellations has taken off yet. It is a bit too early to properly assess how big is the market and what can be done through ridesharing on bigger launchers. Is the market for micro launchers big enough? Are the potential customers ready to pay the right price for such a service? We are open to everything that is happening in the market. We are studying everything very carefully but we only move once we are convinced that there is a business case. For the time being, we are focusing on Ariane 6. We believe that the flexibility of Ariane 6 does address the needs of the small satellite market." Quoted from Pultarova (2017a).

Such advantages are, however, offset by a higher price as compared to other launch solutions, including orbital capsule dispensers, rideshare as secondary payload, and even rideshare. This means that interest in micro launchers will not be equally shared among different customers. As mentioned previously, it is clear that, for example, university CubeSats will continue to prefer piggyback solutions to a dedicated launch.

More broadly, when benchmarking customer requirements with the different launch solutions, it becomes clear that heavier launchers will be better suited for science and education missions as well as constellation deployment and large-scale replenishment of constellations of minisatellites. Conversely, micro launchers will come into play for deploying CubeSat-sized systems, replacing individual mini- or microsatellites in constellations and other single missions not part of bigger systems, such as technology demonstrators, and government (particularly small defence satellites). Micro launcher demand will hence be mainly driven by either institutional actors (particularly the military establishments) or commercial companies (in the telecom and EO markets) seeking tailored services (e.g. assured schedule and availability and high flexibility in orbital delivery) and ready to pay the associated price.

The market that European micro launcher businesses can contemplate is also influenced by a series of factors imposing or encouraging the selection of domestic launch solutions by customers. Factors leading to captive launch markets include legal and regulatory frameworks imposing the selection of domestic solutions, in particular for governmental payloads, but also factors inducing the selection of domestic solutions on presumably open commercial markets (location, customer/supplier relations, etc.). The availability of domestic launch solutions, which explains partially the selection of foreign solutions by European customers, is also an important factor. Overall, the dynamic leading to captive markets is rather complex but, as illustrated in Chart 4.1, national actors tend to select domestic solutions.

As a result, European micro launcher businesses will likely face a series of market barriers that could lead to serious difficulties in capturing foreign small satellites launch markets, including on presumably open commercial markets. By corollary, the demand from European customers, both governmental and commercial, will be an essential driver of the success of European micro launcher projects.

European demand represented around 8% of global demand between 2013 and 2016 in terms of the number of payloads (53 European small satellites), and 12% in terms of mass (3800 kg). Over this period, demand remained relatively stable when compared to the growth observed in Japan, China, and of course, in the USA. During the peak year, in 2013, 23 European small satellites were launched, an exceptional year far above the 15 payloads of 2014 and 7/8 payloads of 2015/2016 (see Chart 4.2).

Historical European demand alone hardly supports a business case for the development of micro launchers (see Chart 4.3):

- In the upper segment of mini- and microsatellites, 2013 and 2014 were the most favourable years for a potential European micro launcher with a capacity around 500 kg. In 2013, 4 payloads were launched for ESA: Swarm A, B and C (468 kg each—launched in cluster with a Rockot-KM) and PROBA V (158 kg—launched

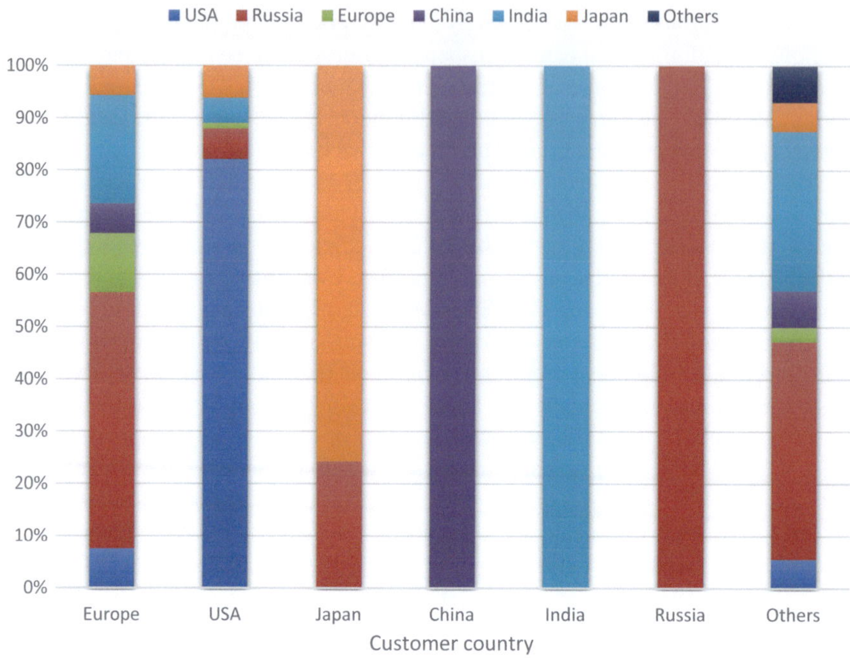

Chart 4.1 Share of satellites launched by country (vertical) organised by customer country (horizontal) between 2013 and 2016. *Source* ESPI

with a foreign 115 kg payload using Vega). In 2015, 4 payloads were launched for UK-based DMC International Imaging and Surrey Satellite Technology (SSTL): DMC-3 1 to 3 (447 kg) and Carbonite 1 (91 kg) launched together on-board a PSLV-XL. For these payloads the suitability of a micro launcher-based solution is not obvious.

- In the lower segment of nanosatellites and CubeSats, although 18 of these satellites were launched in 2013 for European universities and other operators, the trend did not continue with only 11, 3 and 6 European nanosatellites launched in 2014, 2015 and 2016 respectively.

While European demand is projected to expand in the coming years, the majority of missions will continue to be represented by university and education CubeSats. Therefore, with a rather limited institutional (captive) market, European micro launchers will have more limited opportunities than their American counterparts to build a strong business upon institutional anchor customers like the DoD or NASA. Consequently, the success of a European micro launcher will strongly depend on its capacity to capture commercial markets. In certain respects, a parallel can be established with the early days of exploitation of the Ariane launcher, with Europe striving

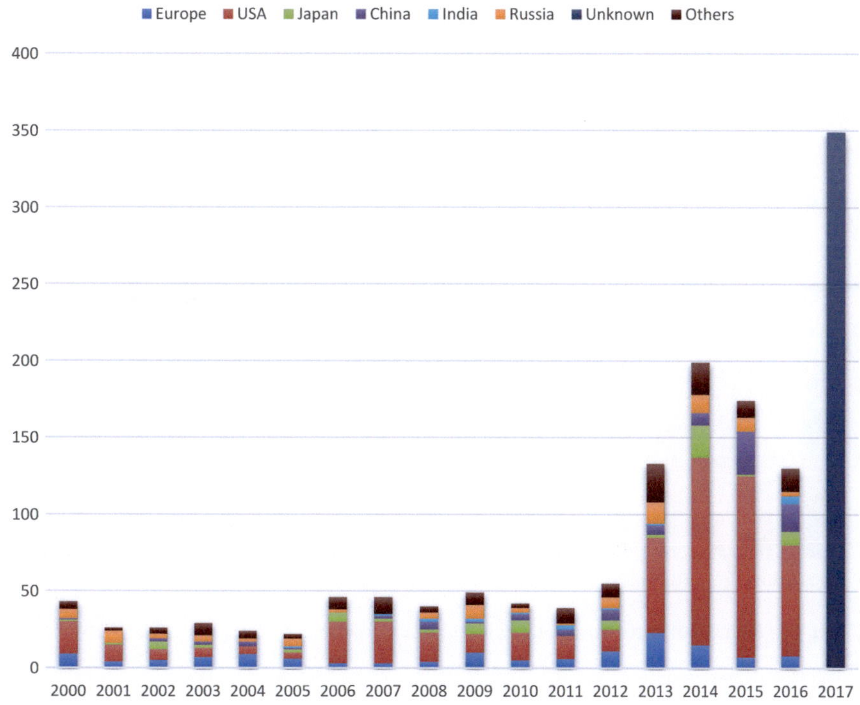

Chart 4.2 Number of small satellites launched by customer country. Detailed analysis not available for 2017 data. *Source* ESPI

to balance both commercial and institutional levers.[25] The dependence of European micro launchers on commercial markets will also dangerously expose their business case to the uncertainty of the materialisation of expected markets (e.g. US-based megaconstellations).

Last but not least, European projects will lead to an operational capacity before 2020, up to two years after the most advanced foreign projects (i.e. Rocket Lab, Virgin Orbit). This late time-to-market is likely to create additional difficulties in positioning a European solution on a market in which competitors are already positioned and are building on experience, launch heritage and established relationships with customers. While this does not constitute a game stopper, it is an incentive for Europe to leapfrog ahead.

Overall, looking more thoroughly into the business perspectives for European micro launchers from the demand side, additional elements come to the fore, eventually contributing to casting more doubts on the solidity of the business case. Table 4.2

---

[25] By now, however, Ariane addresses a mature and quite homogeneous market on which the launcher is extremely well positioned with a value proposition clearly understood and appreciated by customers and an established and continuous demand.

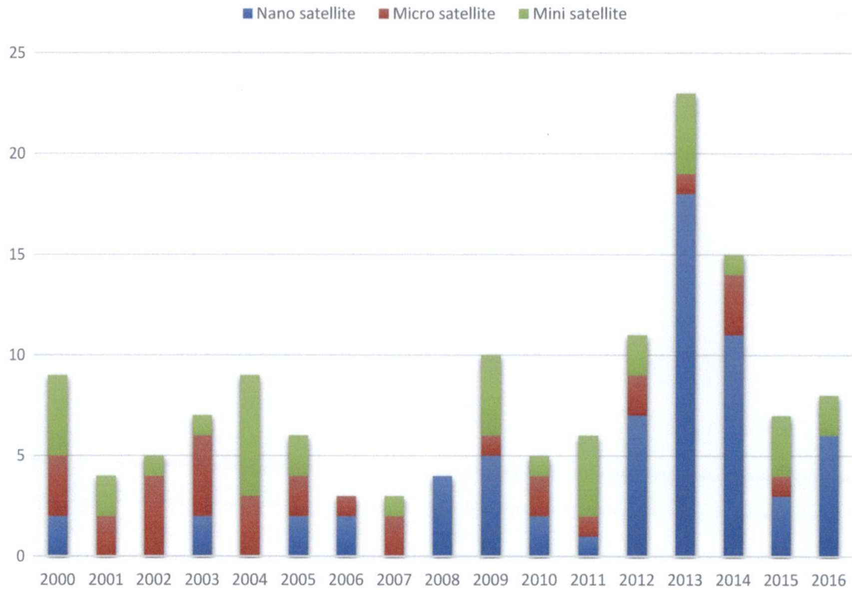

Chart 4.3 Number of European small satellites launched per year by category (2000–2016). *Source* ESPI

Table 4.2 Outstanding issues in the business case of European micro launcher companies

Unknown factors	Known factors
Actual size of the future small satellite market	Fierce competition from existing launch services solutions and their evolutions
Share of the small satellite market possibly intercepted by micro launchers	Traditional launch service solutions will be preferred for mini- and microsatellites constellation deployment, and piggyback for low-budget education missions
Readiness of small satellite operators to choose micro launchers	Commercial demand limited to single missions, CubeSats systems and/or replenishment
Cost competitiveness of the launch price of micro launcher services	Low level of institutional (captive) demand in Europe
Emergence of new classes of missions based on smallsat architecture (Deep Space Science, SSA, M2M for IoT, etc.)	Lack of a European anchor customer

summarises known and unknown factors affecting the business case of European micro launcher companies.

From this standpoint, reinforcing European demand will be an essential element to support the rationale for a European micro launcher. As emerged from the interviews conducted for this research with European small satellite companies and start-ups, there is a perceived lack of support from European Institutions for small satellite operators at this stage: among other issues raised in the interviews, for example, it was noted that the European Investment Bank (EIB) is not allowed to finance dual-use technologies, i.e. most satellite technologies; that the defence market, which could be a potential prime customer for quick and responsive access to space is scattered among different national projects[26]; and that Europe does not (yet?) have the kind of public defence projects that would support such developments (e.g. DIUx in the USA). Small satellite operators and start-ups thus bemoan a lack of incentivisation from pan-European entities for small European commercial smallsat enterprises.

4.1.3 Assessing the Rationale for Micro Launchers in Europe

As clearly emerges from the assessment of the internal demand and supply conditions in the micro launcher market, European initiatives in the sector suffer from several issues affecting the business case.[27] Indeed, one of the dangers is related to the fact that a large share of the global smallsat booming market is taking place in the USA, China and Japan, casting doubts on the number of missions that could effectively be launched by European vehicles.

The lack of a sound commercial business case, however, cannot be automatically taken to imply that there is no need for micro launcher solutions in Europe.

While it is true that the current wave of micro launchers is essentially private-led, embedded in the NewSpace dynamics of the 2010s and focused on the commercial market, it should not be forgotten that the development of any launch vehicle has hitherto responded to both political and economic considerations. To be consistent, the development of micro launchers should be assessed not only from a purely commercial perspective, but also from an institutional one.

If taken in isolation, these two perspectives do not seem to provide solid rationales for the deployment of micro launcher solutions in Europe. However, by integrating and expanding the two different perspectives, a series of substantial stakes eventually come to the fore. This is, *inter alia*, corroborated by the fact that no existing space launcher in history has been able to sustain itself purely through the commercial market. While the situation may gradually change in the future, and in particular as a consequence of the NewSpace dynamics, governments still play the role of key anchor consumers or indirect sponsors of any launch solutions.

- From an institutional perspective, micro launchers are bound to play an increasingly relevant role, for government, civilian and future military space activities.

[26]It must be noted that a military-specific market demand for small satellites does not exist in Europe at the moment. However, considering the global trends in this regard outlined in Sect. 2.3.2, and European-wide efforts to achieve greater efficiency, optimization and integration across different European countries' military structures, also an increased use of small satellites for defence projects across Europe can be considered a likely future trend.

[27]Indeed, one could argue that also no previous or existing launch vehicles were able to sustain itself on the purely commercial market.

As already discussed in Chap. 2.3.2, space assets have become critical infrastructure for a number of key military purposes (e.g. communication, intelligence, early warning). This institutional/military utility clearly explains why all the major spacefaring nations (USA and China first and foremost) have shown strong interest in the deployment of micro launchers despite their uncertain business case and also provides European stakeholders with some important drivers in ensuring a degree of institutional backing and in keeping up to speed with these developments. Even though European military demand is certainly lower (first regarding standard-sized military satellites, and even more so in the small satellites segment) than in the USA, China or Russia, the inherent advantages offered by micro launchers cannot be simply weighted on a typical demand curve. Furthermore, launcher redundancy through micro launcher's solutions could offer—however limited—back-up possibilities in case of the unavailability of European traditional launch solutions, and thus increases their value from a commercial perspective too.

- These considerations on the institutional side show the necessity of broadening the perspective of the assessment also on the commercial side. Two important points need to be highlighted. First, historical analysis of how European smallsats have been launched in recent years highlights the fact that a consistent number of these spacecraft were launched by American, Chinese and Indian vehicles. Second, while the current and projected demand across Europe does not prove per se sufficient to justify the development of micro launcher solutions, one should also recognise that the availability of micro launchers could stimulate the emergence of a supply-driven market, meaning that the increasing supply capacities could stimulate growth in demand, particularly if competition among different launch service providers creates pressures driving down launch costs. Indeed, the fact that micro launchers will have to compete with other launch services solutions will expectably generate a tendency to further lower the price level. It could be even argued that fierce competition among different launch services solutions could eventually induce demand to become more elastic.

As in the case of the institutional drivers, these combined considerations can provide Europe with a stronger case for keeping up with worldwide developments. As of February 2018, European initiatives were not at the same level of preparedness as some of the worldwide advances, particularly in the USA. Furthermore, what is perhaps missing is a degree of institutional clarity among pan-European institutions on the type of steps that could or should be undertaken vis-à-vis micro launcher initiatives.

Accordingly, the following section will provide a set of policy considerations for European stakeholders, be they at national or pan-European level.

4.2 Policy Considerations for Europe

From all the considerations above, we can reasonably conclude that micro launcher vehicles do retain merit when considered from a joint commercial and institutional rationale. Furthermore, from a strategic standpoint, it is clear that Europe is not at the forefront of the multiple current ongoing developments worldwide.

A key issue then is to devise the kind of steps that can be taken to ensure a sufficient level of readiness of European industry in this area and, more specifically, whether there is a role for public institutions in this regard.

4.2.1 Micro Launchers as an Institutional Programme?

The development of a dedicated micro launcher vehicle could be envisioned as part as an ESA optional programme, alongside the current line of launchers, Vega and Ariane, and benefitting from decades of expertise and industrial excellence in European launcher development.

When looking at the upcoming family of European launchers, with Ariane 6.2 and 6.4 covering the medium-to-heavy segment and Vega-C (and Vega-E) covering the medium-to-small part of the performance spectrum, an evolution—or, rather, a scale-down—of Vega in the microdomain could be seen as a natural continuation of ESA-led efforts in European space transportation. The cancelled Sparrow project from ArianeGroup, as well as the announcement of an R&D concept for a Vega Light from Avio, clearly indicates that the two major European industrial actors in space transportation see micro launcher solutions as relevant and of interest.

Overall, such a development would clearly fit into the technical capabilities of the intergovernmental agency. However, additional non-technical considerations should be weighted with respect to such institutional involvement.

First and foremost, before envisioning direct public investment in a micro launcher, it should be assessed whether domestic/regional demand for small satellites is strong enough to support and justify such investment. Indeed, a comprehensive Europe-wide small satellite policy would be a prerequisite for the elaboration of a consistent micro launcher policy.

Second, micro launchers have not yet fully demonstrated their commercial worthiness. As of early 2018, not only has no micro launcher developed in the last wave of initiatives embedded in the NewSpace economy yet achieved a fully commercial launch, but even the most advanced projects from Rocket Lab and Vector—which have already formally signed a number of launch contracts—still have to demonstrate their economic viability. At the moment, many US institutional actors are expressing doubts as to the viability of more than a couple of US providers in this segment of the market, and indeed, various studies conducted by ESA and CNES in recent years have not identified a clear business case for a commercial European micro launcher in the European context so far.

A third consideration is the need to obtain consistent support for such a programme from ESA Member States in an intergovernmental framework, having in mind that several projects—as of yet embryonic—are currently under development on national grounds, notably in Italy, France, Germany, Norway, Spain, Portugal, Sweden and in the United Kingdom (see Sect. 4.1.1).

A fourth reflection, which emerged during interviews conducted with industrial stakeholders, is that direct public involvement in the development phase raises concerns about the inherent complexity of industrial procurement policies of pan-European organisations, which might adversely impact the level of competitiveness, which is key to the business case of micro launchers. In fact, launch price is clearly the most critical factor conditioning the commercial viability of a micro launcher vehicle, and all industry stakeholders interviewed clearly recommended letting market competition unfold by itself. From this standpoint, before considering institutional involvement in the development of European micro launchers, it might be necessary to come up with innovative public procurement schemes aimed at minimising their impact on competitiveness in the exploitation phase.

The potential interference of public endeavours in mostly privately funded markets should also be considered with attention. For example, US private micro launcher developers are raising strong concerns regarding the reuse by the US government of surplus ICBM motors for dedicated micro launchers that they consider as unfair competition. In this respect, European stakeholders should carefully assess to what extent direct institutional involvement in micro launcher vehicles (to be exploited on the global market) could be legally challenged on the international market.[28]

Last but not least, it is clear that Vega in its current version—as well as potential future evolution—and the future Ariane 6 are already targeting the smallsat launch market with appropriate dispensers to accommodate and launch multiple small payloads.[29] The extent to which the development of an institutionally backed lighter version of Vega (or, more broadly, any other European micro launcher) exclusively dedicated to small satellites would adversely affect the exploitation of the larger version of Ariane 6 and Vega-C remains an open question.

The aforementioned considerations revolve around the possibility of *direct* involvement of public agencies in micro launcher development. However, we could also consider different types of *indirect* support from European public institutions to support the emergence of European initiatives and, most importantly, effectively sustain them in the long term. These considerations are discussed in the following paragraph.

---

[28] As for the USA, ICBM-derived micro launchers could be exclusively employed for institutional, national security and defence missions, as it is the case already for Minotaur 1. However, if enacted, a policy of this kind would ultimately subtract a sizeable market share of small satellite missions from private providers.

[29] See ESA's SSMS initiative, (European Space Agency, 2017b).

4.2.2 Institutional Support for the Ecosystem Around Micro Launchers

Indirect support from European institutions aimed at facilitating the emergence of European micro launchers and sustaining their business case could take shape in the ecosystem around micro launcher initiatives. These policy considerations are based on the authors' analyses and are further corroborated by the opinions gathered through interviews with stakeholders.

This type of support can be categorised under four broad kinds of actions, which could be implemented at a coordinated national and/or pan-European level in order to increase the readiness of Europe in the micro launcher segment.

1. Support smallsat developers and development;
2. Support access to space infrastructures;
3. Support the development of KETs;
4. Establish a favourable regulatory and financing framework.

1. Support to the European Small Satellite Market

A possible way forward for pan-European institutions to support the emergence and commercial sustainability of a European micro launcher is to strengthen the demand for such services, with regard to both the institutional and commercial markets.

It is clear that, like the USA, also Europe has taken several initiatives to support smallsats development, and CubeSats in particular. To mention a few, projects funded by pan-European institutions, such as ESA's "Fly your satellite!" and the EU's "QB50" projects,[30] clearly focus on supporting the development and launch of scientific and educational missions.

However, these initiatives, while worthy and commendable for their long-term goal of strengthening Europe's space workforce in the case of "Fly your satellite!", do not have the breadth—nor, indeed, the objective—to support a consistent demand for launch businesses. This is even clearer when considering that, as discussed in the previous chapter, small scientific and educational satellites can hardly constitute the business case for this class of vehicles.[31]

In this respect, the key question is to determine whether—and how—pan-European institutions can support and strengthen the emergence of a commercial (and institutional) demand for smallsats missions. Then, equally relevant, to what extent would and could these measures support micro launcher vehicles without adversely affecting the commercial exploitation of current or upcoming launch solutions from Vega.

[30]It should be mentioned that QB50 participants are not only represented by European institutions, but also universities teams around the world.

[31]Both projects were started when micro launcher vehicles were far from being a concrete launch option, and indeed QB50's CubeSats reached orbit through deployment from NanoRacks and a few PSLV piggyback launches.

In this regard, the NASA Launch Services Programme (LSP) is an interesting example. In 2016, it funded three Venture Class Launch Services contracts to provide small satellites access to LEO,[32] which were awarded at fixed-price conditions worth around $17 m to three micro launcher companies, namely Firefly, Rocket Lab and Virgin Orbit. The aim of the project was "*to foster commercial launch services dedicated to transporting smaller payloads into orbit as an alternative to the rideshare approach and to promote the continued development of the US commercial space transportation industry*", according to Jim Norman, director of Launch Services at NASA.[33]

European actors could likewise initiate mechanisms to sponsor launch opportunities, targeting indigenous smallsat developers and local vehicles at the same time. Other initiatives could be undertaken to spur and support the development of smallsats at component level (as opposed to system level). These could fit within a comprehensive small sat policy, which would in the long-term sustain the case for a European small satellite base—and hence a customer base for micro launcher dedicated access to space.

2. Support to Infrastructure

Ground infrastructures supporting orbital launch are obviously a key element of any space transportation system. At the same time, the operation of a spaceport accounts for a substantial part of the cost of launch exploitation, with typically large recurrent and fixed costs.

In the context of micro launchers, since one of the main selling points of this class of rocket is its availability and quick access to orbit, the existence of an easily accessible launch site to which small satellites can be speedily delivered, integrated and launched is a prerequisite for the success of any European micro launcher initiative.

Naturally, the Guyana Space Centre, the European gateway to space located in Kourou could, in principle, be considered as the prime location to host and launch either private or public-led European micro launchers. With its fully operational infrastructures already in place for the exploitation of Ariane and Vega, an expansion could be envisioned to further accommodate micro launchers; even more so if considering that its fixed costs (e.g. radar, telemetry, security, spacecraft processing.) are already paid for by the activities of the European family of launchers. However, during interviews with micro launcher and smallsat stakeholders, it emerged that Kourou could be impractical for this type of business, owing to considerations related to cost- and time-effectiveness (not least, an inherent difference in the type of business between large and small satellites).

Furthermore, micro launcher companies select a spaceport to launch from chiefly by virtue of its location—among other safety, regulatory and licensing considerations—since the latitude has an impact on orbital launch parameters (and thereby the targeted smallsat market). From a micro launcher point of view, being in a way "spaceport-agnostic" is key to addressing the full scope of the market.

---

[32]National Aeronautics and Space Administration (2015a).

[33]Ibid.

From this standpoint, as mentioned in Sect. 2.4.3, it is highly relevant that a number of European countries are developing new spaceports or upgrading existing suborbital launch infrastructures, to eventually accommodate the launch of emerging micro launchers from European continental soil. Furthermore, in several instances, the institutional backing for such ground infrastructures is often accompanied by funding for the development of indigenous micro launcher technology and vehicles.

The various initiatives at national level are certainly to be recognised and commended. However, when addressing the topic from the pan-European worthiness perspective, further steps could be taken with the ultimate objective of supporting a strategic infrastructure for all-spectrum European independent access to space. In this sense, the involvement of pan-European public institutions on top of national spaceports-related initiatives could be considered as part of a policy of indirect support for micro launchers.

Such support would not only be instrumental in easing the pressure on micro launcher businesses to close their business case, but would also ensure that European micro launchers are backed by European top-grade access to space infrastructures, thereby increasing investor and customer confidence in funding and selecting European micro launchers.

3. Support for the Development of KETs

Regardless of the potentially—successful or unsuccessful—commercialisation of micro launcher solutions, the dynamism of the international launch industry and the increasingly fierce global competition—not only among different providers but also among different typologies of launch solutions (piggybacking, rideshares, etc.)—pose challenges that European stakeholders cannot avoid taking up. Such challenges dictate a steady and constant investment in the development of launchers-related Key Enabling Technologies (KETs) and, more broadly, a strong commitment to the pursuit of disruptive technologies that can substantially lower the cost of access to space, thus actively supporting the competitiveness of the European launcher sector.

Currently, the major launcher-related R&D efforts are driven and supported by ESA through its Future Launcher Preparatory Programme (FLPP). The FLPP is an optional programme for development and maturation of technologies to be used in future European launch vehicles. Initiated in 2003, the programme's initial aim was to develop technologies for the successor of Ariane 5. With the inception of the Ariane 6 project, the focus of FLPP was shifted to a general development of new technologies for European launchers. The stated objectives of the FLPP are to:

- Identify and prepare the system competence and technology for development with the aim of confining launcher time-to-market within 5 years, reducing recurring cost and development risk, while keeping long-term industry competitiveness;
- Promote reusability of existing and new technologies to reduce development costs globally;

- Perform system studies to assess evolutions of operational launchers, future launcher architectures, advanced concepts, select technology and elaborate technology requirements;
- Safeguard critical European industrial capabilities for the safe exploitation of the current launchers and guaranteed access to space;
- Develop environmentally friendly technologies.[34]

FLPP hence develops and matures technologies that, though promising for future application, do not have a sufficient Technology Readiness Level (TRL) to allow them to be integrated into development programmes.[35] Key technology areas financed by the FLPP include:[36]

- Lightweight and high-performance systems;
- Spin-in and use of commercial off-the-shelf technologies;
- More electric-based launchers;
- Mastering the environments of launch;
- Orbit injection strategies;
- Robust structures and design;
- Ease of manufacture, operation and integration;
- Reusability of existing and new technologies;
- Low-cost structures and systems;
- Green launch systems.

Given the importance of this type of programme but the rather limited contributions it receives, considerations could be given to how to financially expand the size of the FLPP, or to whether additional actors could potentially come into play via specific measures. Indeed, outside the ESA framework, the most important pan-European instrument for pursuing technological innovation in general is the EU Framework Programmes for Research and Technological Development.

In this respect, it could be argued that pursuing disruptive solutions in the high technology segment is a matter where not only ESA but also the EU should consider taking more substantial action. As pointed out in a previous ESPI study, two main reasons supporting complementary efforts can be pinpointed:

- First, the EC is generally better positioned to initiate programmes that pursue disruptive innovation. As opposed to sustaining innovation, the outcome of disruptive innovation is not guaranteed and thus it entails a stronger risk and uncertainty in terms of pay-offs. In this sense, this type of innovation requires strong political commitment and support—something that can more easily be provided by the EC, which in principle has a more risk-tolerant mandate.

---

[34]European Space Agency (2017a).

[35]Those technologies typically have an initial TRL of 3 or lower and the objective is to raise the TRL up to about 6.

[36]European Space Agency (2017a).

- Second, the EU as an actor has an interrelated number of interests and tools to support disruptive innovation in terms of industrial policy, promotion of European industrial competitiveness, and research and innovation schemes.[37]

Possible EU efforts towards ensuring the technological preparedness of European launch capabilities over the long term (be it on the large or the small side of the mass spectrum) should thus be seen as complementary—not alternative—to those of ESA. Considerations could be ideally given to the possibility of devoting an amount of seed funding within the next Multiannual Financial Framework (MFF) programme to the development of KETs. Indeed, several calls covering the topic of "independent access to space" and "bottom-up technologies at low TRL" for industrial competitiveness have already been approved over the last few years.[38]

Ideally, additional funding could thus be allocated to develop other KETs in the sector, so as to enable Europe to maintain a technological competitive advantage vis-à-vis other players, while also allowing established and new European companies involved in the launcher sector to reap the benefits (primarily costs savings) of new generations of launchers.

Irrespective of the specific technological development to be financed, what needs to be more broadly highlighted is that these possible measures to lower the cost of access to space by means of technological development may in turn also contribute to making future micro launch system offerings more competitive—and hence sustainable—from a commercial perspective.

4. Support to the Financing and Regulatory Framework

A final type of measure that might positively impact on micro launcher initiatives lies in the introduction of financing and regulatory mechanisms intended to support the growth of an investment ecosystem in Europe.

To be sure, Europe is already one of the biggest ecosystems for entrepreneurs.[39] Thanks in particular to such ESA-led initiatives as the Technology Transfer Programme (TTP) and Business Incubation Centres (BICs) not only can Europe today boast a large number of start-ups being launched every year, but also an increasingly large network of brokers, business angels, incubators, mentors and accelerators, as well as universities, research centres and experts (see Fig. 4.1).

What is still lacking in Europe is a dedicated space finance community providing equity finance vehicles for space entrepreneurs. This might appear odd considering the massive liquidity in financial markets, but the reality is that the investment level is still cumbered by a number of factors, including: the "lack of familiarity with the sector and associated market opportunities, difficulties associated with performing due diligence on space start-ups", and the lack of effective exit strategies that can win

---

[37] Aliberti and Tugnoli (2016, p. 80).

[38] Horizon 2020 Calls for research into breakthrough technologies to provide access to space and for research into Key Enabling Technologies with relevance for the fields of energy production, energy storage, material and structures, additive layer manufacturing techniques, etc.

[39] Kreisel (2017).

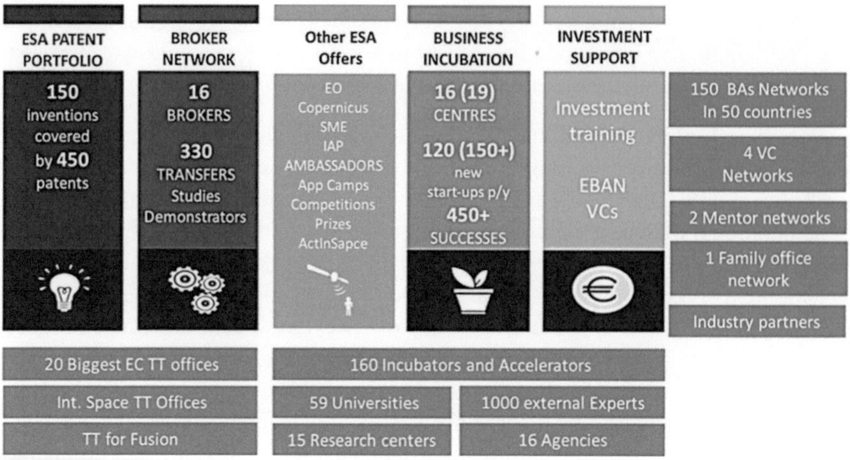

Fig. 4.1 Europe's Space business ecosystem. *Source* Kreisel (2017)

investors' confidence.[40] This inevitably impacts the quality and quantity of investment opportunities in the space sectors business, including in the micro launchers segment.

Given this background, dedicated initiatives aimed at encouraging investment in seed- and early-stage space businesses could ideally be introduced by the EC, ESA and national governments, together with innovation incentive prizes as well as an improved regulatory framework that is supportive of space entrepreneurship at the policy level. An example is the "Low-Cost Space Launch" EIC Horizon Prize, funded by the EC with an amount of €10 m, aiming at "*developing a solution for launching satellites into low orbit at a low cost*".[41] This initiative is expected to start in the first quarter of 2018, with the prize to be awarded in 2021.

As recognised and commented on by many interviewees in the framework of this report, as well as by the OECD in its 2016 *Space and Innovation* report, the establishment of a comprehensive regulatory, legal and procedural regime for conducting private space activities,[42] and the enactment of pro-business policy mechanisms (i.e. grants, procurement mechanisms, prizes), figure among the most important tools for encouraging entrepreneurial activity, and more broadly innovation, in the space sector.[43] These regulations need to provide as much clarity as possible with respect,

---

[40]Ibid.

[41]European Commission (2017).

[42]These regulations need to provide as much clarity as possible with respect for instance to licensing, technology certification, insurance, liability and dispute resolutions among the others.

[43]Organisation for Economic Co-operation and Development (2016). In the analysis of the apposite conditions for innovation, the OECD more specifically identified: (a) an efficient system for knowledge creation and diffusion in institutions (from fundamental knowledge to technology transfer to other sectors); (b) policies that encourage innovation and entrepreneurial activity, while providing as much clarification as possible; (c) business environment that encourages investment in technology

for instance, to licensing, technology certification, insurance, liability and dispute resolutions among others. In short, what is needed at the policy level is to ensure that the regulatory environment associated with launches (licences, Space Acts, etc.) makes Europe a good place to do business.

All in all, it appears evident that by creating an appropriate investment and regulatory framework for the broader development of "astropreneurship" in Europe, European institutions could indirectly also help micro launcher companies' move towards steadier and more sustainable growth.

4.3 Conclusions

Over the past few years, the proliferation of numerous micro launcher initiatives worldwide has started to raise questions about the existence of a dedicated institutional or commercial market for micro launchers. This report has undertaken a comprehensive assessment of the structure and dynamics of such markets, focussing particularly on the European context. Relevant conclusions are as follows.

- Uncertainties around the future structure of the small satellite market, and even more so around the business case for micro launchers, persist. However, while the CubeSats and nanosatellite segment is in the process of being structured (one of the reasons being the now-widely adopted standard of CubeSats), and covered by a number of already available launch solutions, the upper segment of mini- and microsatellites is seemingly more open to competition for accessing space and for micro launchers in particular.
- There is not one single sizeable business case for micro launcher vehicles, but rather a diverse ensemble, which includes satellites of different classes of mass, mission types and customer origin—and micro launchers themselves equally come in a variety of sizes and performances. Therefore, companies will necessarily need to aggregate a very heterogeneous demand to ensure successful exploitation.

---

and in knowledge-based capital (allowing experiment with new ideas, technologies and business models); (d) a skilled workforce.

The OECD also highlighted three high-level policy responses to better monitor and encourage space innovation. According to the OECD, national governments should more specifically: (a) evaluate national policy instruments that support space innovation, with particular attention to examining the networks of knowledge diffusion, such as clusters and incubators, to ensure complementarity at regional and national levels; (b) map downstream space activities: there are pressures on governments to adopt reforms that enable start-ups and innovative firms to find or to retain niches in which they may make the most of their capabilities. This includes mapping actors along the value chains; (c) capture and document technology transfers & commercialisation (TTC), because a significant outcome from government-funded space research consists of space technology transfers leading to the development of new commercial products and services in various economic sectors (e.g. transport, health, environment), and the creation of spin-off companies. (i.e. Space Agencies Technology Transfer Officers group). In this respect, there is a still a long way to systemise the various data in a more coherent fashion.

- In the European Strategy on Space, autonomous access to space is identified among Europe's strategic priorities. The question arises as to whether this would extend also to the segment represented by micro launchers. However, the fact that these vehicles do not constitute a critical means for accessing space (despite their potential value offer), coupled with a European demand (institutional first, then commercial) that so far is very limited, suggests that they do not constitute a compelling strategic requirement for ensuring Europe's independent access to space. At the same time, when looking at worldwide developments and future trends, it can be argued that institutions should be ready to adjust their stance in this regard, either through direct or indirect means as outlined above.
- Low historic demand and so far limited domestic markets for small satellites in Europe suggest the need for a thorough assessment of the projections, potential and evolution of the European small satellites market, in particular in light of skyrocketing capabilities that are today being exploited by commercial companies, but could be equally relevant for institutional actors in years to come. In this regard, identifying or encouraging the establishment of a European small satellite "anchor customer" would be strongly beneficial for the long-term sustainability of any European micro launcher initiative.
- Indeed, a more comprehensive and encompassing European small satellite policy (in terms of technologic standards and components development, strategic goals, regulatory frameworks, R&D support, etc.) should precede any strong institutional involvement in launcher development. This comprehensive policy should furthermore include continental spaceports in its scope.
- Direct involvement of European institutions in micro launchers to be then exploited on the global market needs to be thoroughly assessed, also from a governance and legal perspective. On the former point, considerations should be given to the optimal framework for the commercialisation of a possible micro launcher vehicle within the European setting: it remains to be seen whether possible micro launcher development would include or exclude Arianespace as a potential dealer on the commercial market. On the latter point, consideration should be paid to the risk of being challenged for unfair competition by purely privately funded competitors on the global market.

List of Acronyms

A	
AFB	Air Force Base
AG	ArianeGroup
AIS	Automatic identification system
ALASA	Airborne Launch Assist Space Access
ASI	Agenzia Spaziale Italiana
ASLV	Augmented/Advanced Satellite Launch Vehicle
ASRI	Australian Space Research Institute
AVUM	Attitude Vernier Upper Module
B	
B2B	Business to Business
B2G	Business to Government
BIC	Business Incubation Centre
BLOS	Beyond Line-of-Sight
C	
CAD	Computer-aided design
CASIC	China Aerospace Science and Industry Corporation
CEO	Chief executive officer
CFO	Chief financial officer
CGWIC	China Great Wall Industry Corporation
CNES	Centre National d'Etudes Spatiales
CONAE	Comisión Nacional de Actividades Espaciales
COTS	Commercial off-the-shelf
CSLA	Commercial Space Launch Act
CTO	Chief technology officer
CYGNSS	Cyclone Global Navigation Satellite System
D	
DARPA	Defense Advanced Research Projects Agency

(continued)

M. Tugnoli et al., *European Access to Space: Business and Policy Perspectives on Micro Launchers*, SpringerBriefs from the European Space Policy Institute
https://doi.org/10.1007/978-3-319-78960-6

(continued)

DIUx	Defense Innovation Unit Experimental
DLR	Deutschen Zentrums für Luft- und Raumfahrt
DoD	Department of Defense
E	
EC	European Commission
EIB	European Investment Bank
EIC	European Innovation Council
ELV	European Launch Vehicle
ENAC	Ente Nazionale Aviazione Civile
EO	Earth observation
ESA	European Space Agency
ESPI	European Space Policy Institute
EU	European Union
EUMETSAT	European Organisation for the Exploitation of Meteorological Satellites
F	
FAA	Federal Aviation Administration
FLPP	Future Launchers Preparatory Programme
G	
GEO	Geostationary orbit
I	
IAE	Institute of Aeronautics and Space
ICBM	Intercontinental Ballistic Missile
ICT	Information and communication technology
IGS	Innovation and growth strategy
ILS	International Launch Services
INTA	Instituto Nacional de Técnica Aeroespacial
IoT	Internet-of-Things
ISA	Iranian Space Agency
ISRO	Indian Space Research Organisation
ISS	International Space Station
ITAR	International Traffic in Arms Regulations
ITU	International Telecommunication Union
J	
JAXA	Japan Aerospace Exploration Agency
K	
KET	Key Enabling Technologies
L	
LEO	Low earth orbit
LLL	Low-cost lightsat launch
LSP	Launch service programme
M	
M2M	Machine to machine

(continued)

(continued)

MDA	Missile Defence Agency
MFF	Multiannual Financial Framework
MLA	Airborne Micro Launcher
MLS	Microsat Launch Share
MTCR	Missile Technology Control Regime
N	
NASA	National Aeronautics and Space Administration
NEA-Scout	Near-Earth Asteroid Scout
NGSO	Non-Geostationary Satellite Orbit
NLR	Netherlands Aerospace Center
NOAA	National Oceanic and Atmospheric Administration
NRCSD	NanoRacks CubeSat Deployer
NSLV	North Star Launch Vehicle
NSTS	National Space Technology Strategy
O	
OECD	Organisation for Economic Co-operation and Development
OHB	OHB SE
ONERA	Office National d'Etudes et de Recherches Aérospatiales
ORS	Operationally Responsive Space Office
OTS	Off-the-shelf
P	
PSLV	Polar Satellite Launch Vehicle
PTD	Pathfinder Technology Demonstrator
R	
R&D	Research and development
S	
SALVO	Small Air Launch Vehicle to Orbit
SMDC-ONE	Space and Missile Defense Command-Operational Nanosatellite Effect
SME	Small and medium-sized enterprise
SMILE	Small Innovative Launcher for Europe
SSA	Space Situational Awareness
SSMS	Small Satellite Mission Service
SSO	Sun-synchronous orbit
SSTL	Surrey Satellite Technology
SSTP	Small Spacecraft Technology Programme
T	
TRL	Technology Readiness Level
TT&C	Technology Transfers & Commercialisation
TTP	Technology Transfer Programme
U	
U.S.	United States of America

(continued)

(continued)

U.S.S.R.	Union of Soviet Socialist Republics
UK	United Kingdom
UKSA	UK Space Agency
ULA	United Launch Alliance
URSC	United Rocket and Space Corporation
USAF	United States Air Force
V	
VC	Venture capital
VEGA	Vettore Europeo Generazione Avanzata
VLM	Veículo Lançador de Microssatélites
VLS	Veículo Lançador de Satélites
VSS	Vector Space Systems

References

Adamowski, J. (2017, November 21). *Sweden eyes small satellite market with expanded space center*. Retrieved from SpaceNews: http://spacenews.com/sweden-eyes-small-satellite-market-with-expanded-space-center/.

Afilipoae, T.-P., Neculăescu, A.-M., Onel, A.-I., Pricop, M.-V., Marin, A., Perşinaru, A.-G., . . . Chel, T.-V. (2018). Launch Vehicle - MDO in the development of a Microlauncher. *Transportation Research Procedia* (pp. 1-11). Elsevier.

Agenzia Spaziale Italiana. (2016, June 30). *A spaceport in Italy?* Retrieved from Agenzia Spaziale Italiana website: https://www.asi.it/en/news/a-spaceport-in-italy.

Air University. (2017). *Fast space: Leveraging ultra low-cost space access for 21th century challenges.*

Aldrin, A. J. (1998). Technology control regimes and the globalization of space industry. *Space Policy, 14* (2), 115–122.

Aliberti, M., & Tugnoli, M. (2016). *The European launchers between commerce and geopolitics.* Vienna: European Space Policy Institute.

ANSA. (2017, June 21). *Future Vega launcher models showcased.* Retrieved from ANSA website: http://www.ansa.it/english/news/science_tecnology/2017/06/21/future-vega-launcher-models-showcased_ef2693b8-6752-4b91-bd03-c56461ccc07c.html.

ArianeGroup. (2017, June 20). *After a very active first half in 2017, Arianespace signs its initial launch contract for the new Vega C launcher and gears up for Ariane 6.* Retrieved from ArianeSpace website: http://www.arianespace.com/press-release/after-a-very-active-first-half-in-2017-arianespace-signs-its-initial-launch-contract-for-the-new-vega-c-launcher-and-gears-up-for-ariane-6/.

Asia-Pacific Satellite Communications Council Newsletter. (2017). *LEO satellite constellations.* Retrieved from Asia-Pacific Satellite Communications Council website: http://www.apscc.or.kr/upload/pdf/Q2%202017.pdf.

Barna, J. (2015). *Exponential improvements in SmallSat technology.* Retrieved from ITU Small Workshop website: https://www.itu.int/en/ITU-R/space/workshops/2015-prague-small-sat/Presentations/Spire.pdf.

BIS Research. (2017). *Global nano satellite market—Analysis and forecast (2017–2021).* Retrieved from BIS Research website: https://bisresearch.com/industry-report/global-nano-satellite-market-2021.html.

Bryce Space and Technology. (2017). *State of the satellite industry report.* Retrieved from Satellite Industry Association website: https://www.sia.org/wp-content/uploads/2017/07/SIA-SSIR-2017.pdf.

Catapult Satellite Applications. (2017). *Small satellite market intelligence.* Retrieved from Catapult Satellite Applications website: https://sa.catapult.org.uk/services/market-reports/small-satellite-market-intelligence-registered/.

M. Tugnoli et al., *European Access to Space: Business and Policy Perspectives on Micro Launchers*, SpringerBriefs from the European Space Policy Institute
https://doi.org/10.1007/978-3-319-78960-6

Clark, S. (2015, November 4). *Air force declares failure on Super Strypi test launch*. Retrieved from Spaceflight Now website: https://spaceflightnow.com/2015/11/04/air-force-declares-failure-on-super-strypi-test-launch/.

Clark, S. (2016, December 27). *Orbital ATK has no plans to phase out seldom-used Pegasus rocket*. Retrieved from Spaceflight Now: https://spaceflightnow.com/2016/12/27/orbital-atk-has-no-plans-phase-out-seldom-used-pegasus-rocket/.

Clark, S. (2017, January 9). *Kuaizhou rocket lifts off on first commercial mission*. Retrieved from Spaceflight Now website: https://spaceflightnow.com/2017/01/09/kuaizhou-rocket-lifts-off-on-first-commercial-mission/.

Davis, J. (2016, February 10). *SLS CubeSats to set sail for deep space*. Retrieved from The Planetary Society website: http://www.planetary.org/blogs/jason-davis/2016/20160210-sls-cubesats-deep-space.html.

Dedieu, V., & Menetrier, L. (2016). *Open Space, l'ouverture comme réponse aux défis de la filière spatiale*.

de Selding, P. B. (2015a, April 16). *Brazil pulling out of Ukrainian launcher project*. Retrieved from SpaceNews: http://spacenews.com/brazil-pulling-out-of-ukrainian-launcher-project/.

de Selding, P. B. (2015b, January 23). *Signs of a Satellite internet gold rush in burst of ITU filings*. Retrieved from SpaceNews: http://spacenews.com/signs-of-satellite-internet-gold-rush/.

de Selding, P. B. (2015c, June 16). *Launch options were key to Arianespace's OneWeb Win*. Retrieved from SpaceNews: http://spacenews.com/launch-options-were-key-to-arianespaces-oneweb-win/.

eoPortal Directory. (2016). *ISS utilization: NanoRacks logistics services for small satellites and ISS deployment systems*. Retrieved from eoPortal Directory Website: https://directory.eoportal.org/web/eoportal/satellite-missions/i/iss-nanoracks-services.

Euroconsult. (2017, July 13). *$30 billion market value for small satellites over coming decade*. Retrieved from Euroconsult website: http://www.euroconsult-ec.com/13_July_2017.

European Commission. (2016). *SMILE—SMall innovative launcher for Europe*. Retrieved from CORDIS: https://cordis.europa.eu/project/rcn/200829_en.html.

European Commission. (2017). *Low-cost space launch*. Retrieved from EC website: https://ec.europa.eu/research/eic/index.cfm?pg=prizes_space.

European Space Agency. (2017a, May 10). *FLPP preparing for Europe's next-generation launcher*. Retrieved from ESA website: http://www.esa.int/Our_Activities/Space_Transportation/New_Technologies/FLPP_preparing_for_Europe_s_next-generation_launcher.

European Space Agency. (2017b, February 24). *Vega flight opportunity for multiple small satellites*. Retrieved from ESA website: http://www.esa.int/Our_Activities/Space_Transportation/Vega_flight_opportunity_for_multiple_small_satellites.

European Space Agency. (2018, February 8). *ESA explores microlaunchers for small satellites*. Retrieved from ESA website: http://www.esa.int/Our_Activities/Space_Transportation/ESA_explores_microlaunchers_for_small_satellites.

Fought, B. E. (1988). Legal Aspects of the commercialization of space transportation systems. *Berkeley Technology Law Journal, 3*(1), 99–147.

Foust, J. (2015, June 12). *Smallsat developers enjoy growth In launch options*. Retrieved from SpaceNews: http://spacenews.com/smallsat-developers-enjoy-growth-in-launch-options/.

Foust, J. (2016a, November 22). *Big data a big market for small satellites*. Retrieved from SpaceNews: http://spacenews.com/big-data-a-big-market-for-small-satellites/.

Foust, J. (2016b, November 16). *Spire deploys four satellites from Cygnus*. Retrieved from SpaceNews: http://spacenews.com/spire-deploys-four-satellites-from-cygnus/.

Foust, J. (2017a, November 29). *British government to offer funding for spaceports and launchers*. Retrieved from SpaceNews: http://spacenews.com/british-government-to-offer-funding-for-spaceports-and-launchers/.

Foust, J. (2017b, June 20). *Glavkosmos seeks to become a major smallsat launch provider*. Retrieved from SpaceNews: http://spacenews.com/glavkosmos-seeks-to-become-a-major-smallsat-launch-provider/.

Foust, J. (2017c, June 23). *Rideshare demand grows despite development of small launch vehicles*. Retrieved from SpaceNews: http://spacenews.com/rideshare-demand-grows-despite-development-of-small-launch-vehicles/.

Foust, J. (2017d, March 3). *SpaceX delays force Spaceflight to find alternative launches*. Retrieved from SpaceNews: http://spacenews.com/spacex-delays-force-spaceflight-to-find-alternative-launches/.

Foust, J. (2017e, August 3). *Vector performs second test flight of smallsat launch vehicle*. Retrieved from SpaceNews: http://spacenews.com/vector-performs-second-test-flight-of-smallsat-launch-vehicle/.

Foust, J. (2017f, November 16). *Virgin Orbit wins first defense department launch contract*. Retrieved from SpaceNews: http://spacenews.com/virgin-orbit-wins-first-defense-department-launch-contract/.

Foust, J. (2018, January 20). *Rocket Lab Electron reaches orbit on second launch*. Retrieved from SpaceNews: http://spacenews.com/rocket-lab-electron-reaches-orbit-on-second-launch/.

Government of India—Department of Space. (2016, July 20). *Privatisation of PSLV* . Retrieved from Press Information Bureau—Government of India: http://pib.nic.in/newsite/PrintRelease.aspx?relid=147347.

Gruss, M. (2015, November 30). *DARPA scraps plan to launch Small Sats from F-15 fighter jet*. Retrieved from SpaceNews: http://spacenews.com/darpa-airborne-launcher-effort-falters/.

Haemmerli, B., Boiron, A. J., & Verberne, O. (2016). The Norwegian initiative for a satellite nano-launcher. In *29th Annual AIAA/USU Conference on Small Satellites*.

Hancock, R. (2005). Provisions of the Commercial Space Launch Act (CSLA). *Space Policy, 21* (3), 227–229.

Henry, C. (2015, August 20). *New tech: US Military's new space technology ambitions*. Retrieved from Satellite Today Website: http://interactive.satellitetoday.com/new-tech-us-militarys-new-space-technology-ambitions/.

Henry, C. (2017, October 12). *General atomics ramping cubesat production, muses railgun smallsat launcher*. Retrieved from SpaceNews: http://spacenews.com/general-atomics-ramping-CubeSat-production-muses-railgun-smallsat-launcher/.

Krebs, G. D. (2017). *Electron*. Retrieved from Gunter's Space Page website: http://space.skyrocket.de/doc_lau/electron.htm.

Krebs, G. D. (2018). *Scout-G1*. Retrieved from Gunter's Space Page: http://space.skyrocket.de/doc_lau_det/scout-g1.htm.

Kreisel, J. (2017, September 12, 13). *Astropreneurs in Europe*. Retrieved from ESPI website: https://www.espi.or.at/images/documents/11_autumn_conf/Presentations/Kreisel/JKIC-ESPI-AutumConf2017-1-FINAL.pdf.

Lim, D. (2016). *Small launcher market survey*. Retrieved from Room The Space Journal: https://room.eu.com/article/small-launcher-market-survey.

Markets and Markets. (2017). *Small satellite market worth 7.53 billion USD by 2022*. Retrieved from Markets and Markets website: https://www.marketsandmarkets.com/PressReleases/small-satellite.asp.

Martin, H. (2016). *NanoRacks ISS CubeSat deployment*. Retrieved from CubeSat Developers Workshop: http://mstl.atl.calpoly.edu/~bklofas/Presentations/DevelopersWorkshop2016/5_HenryMartin.pdf.

McConnaughey, P. K., Femminineo, M. G., Koelfgen, S. J., Lepsch, R. A., Ryan, R. M., & Taylor, S. A. (2010, November). *DRAFT launch propulsion systems technology Area 01*. Retrieved from NASA website: https://www.nasa.gov/pdf/500393main_TA01-LaunchPropulsion-DRAFT-Nov2010-A.pdf.

Messier, D. (2016, October 3). *A plethora of small satellite launchers*. Retrieved from Parabolic Arc website: http://www.parabolicarc.com/2016/10/03/plethora-small-sat-launchers/.

NASA Ames Research Center. (2015). *Small spacecraft technology state of the art*. NASA.

National Aeronautics and Space Administration. (2014, March 28). *NASA awards launch services contract for CYGNSS mission*. Retrieved from NASA website: https://www.nasa.gov/centers/kennedy/news/releases/2014/release-20140328.html.

National Aeronautics and Space Administration. (2015a, October 14). *NASA awards venture class launch services contracts for CubeSat satellites*. Retrieved from NASA website: https://www.nasa.gov/press-release/nasa-awards-venture-class-launch-services-contracts-for-cubesat-satellites.

National Aeronautics and Space Administration. (2015b, October 14). *Pathfinder technology demonstrator—Demonstrating novel CubeSat technologies in Low Earth Orbit*. Retrieved from NASA Website: https://www.nasa.gov/directorates/spacetech/small_spacecraft/ptd.html.

National Aeronautics and Space Administration. (2015c, February 26). *What are SmallSats and CubeSats?* Retrieved from NASA website: https://www.nasa.gov/content/what-are-smallsats-and-cubesats.

Nordic Investment Bank. (2015, April). *Swedish space corporation: An arctic gateway to space*. Retrieved from Nordic Investment Bank website: https://www.nib.int/news_publications/cases_and_feature_stories/1607/swedish_space_corporation_an_arctic_gateway_to_space.

OHB SE. (2018, February 13). *OHB SE: MT Aerospace is conducting research into new concepts for commercial micro-launchers on behalf of ESA*. Retrieved from Dgap.de website: http://www.dgap.de/dgap/News/corporate/ohb-aerospace-conducting-research-into-new-concepts-for-commercial-microlaunchers-behalf-esa/?newsID=1052587.

Orbex. (2017, July 11). *Orbex wins Horizon 2020 SME Instrument Grant for space launch vehicle tanks*. Retrieved from Orbex Website: http://www.orbex.space/orbex-wins-horizon-2020-sme-instrument-grant-for-space-launch-vehicle-tanks/.

Organisation for Economic Co-operation and Development. (2016). *Space and innovation*. Paris: OECD.

Oving, B., van Kleef, A., Verberne, C. J., Haemmerli, B., Kuhn, M., Müller, I., et al. (2016). Small Innovative Launcher for Europe. In *4S Symposium.* Valletta.

Platzer, P. (2016, April 14). *The bottle-sized satellites leading a quiet revolution in space*. Retrieved from World Economic Forum website: https://www.weforum.org/agenda/2016/04/get-ready-for-the-nanosatellite-revolution/.

PLD Space. (2016, November 7). *PLD Space gets ESA support to develop the first reusable launch vehicle technologies in Europe*. Retrieved from PLD Space website: http://www.pldspace.com/blog/en/2016/11/07/pld-space-esa-support-reusable-launch-vehicle-europe/.

Pultarova, T. (2017a, October 25). *ArianeGroup CFO Pierre Godart on Ariane 6 cost savings, micro launchers and reusability*. Retrieved from SpaceNews: http://spacenews.com/arianegroup-cfo-pierre-godart-on-ariane-6-cost-savings-micro-launchers-and-reusability/.

Pultarova, T. (2017b, September 13). *Big launch companies predict doom for upcoming smallsat launchers*. Retrieved from SpaceNews: http://spacenews.com/big-launch-companies-predict-doom-for-upcoming-smallsat-launchers/.

Richardson, D. (2017, July 14). *Soyuz rocket sends 73 satellites into 3 different orbits*. Retrieved from Spaceflight Insider website: http://www.spaceflightinsider.com/organizations/roscosmos/soyuz-rocket-sends-73-satellites-3-different-orbits/.

Russell, K. (2017, May 4). *SpaceX outlines plans for high-speed broadband mega-constellation*. Retrieved from SatelliteToday website: http://www.satellitetoday.com/telecom/2017/05/04/spacex-outlines-plans-high-speed-broadband-mega-constellation/.

Rycroft, M. J., & Crosby, N. (Eds.). (2002). *Smaller satellites: Bigger business?*. Netherlands: Springer.

Schingler, R. (2017, February 14). *Planet launches satellite constellation to image the whole planet daily*. Retrieved from Planet website: https://www.planet.com/pulse/planet-launches-satellite-constellation-to-image-the-whole-planet-daily/.

Seibert, G. (2006, November). *The history of sounding rockets and their contribution to European Space Research.* Retrieved from ESA website: http://www.esa.int/esapub/hsr/HSR_38.pdf.

Singh, S. (2017, November 22). *Isro's plan: A rocket that can be made in 3 days*. Retrieved from The Times of India website: https://timesofindia.indiatimes.com/india/isros-plan-a-rocket-that-can-be-made-in-3-days/articleshow/61746668.cms.

Space Services Inc of America. (2014). *The launch of Conestoga 1*. Retrieved from Space Services Inc of America website: http://www.spaceservicesinc.com/conestoga-1.

SpaceWorks. (2017, January 30). *SpaceWorks announces release of 2018 nano/microsatellite market forecast*. Retrieved from SpaceWorks website: http://spaceworkseng.com/spaceworks-announces-release-of-2018-nanomicrosatellite-market-forecast/.

Surrey Satellite Technology. (2017). *Satellite data applications—Technology demonstration*. Retrieved from SSTL website: https://www.sstl.co.uk/Satellite-Data-Applications/Technology-Demonstration.

Taverney, T. D. (2017, December 1). *Op-Ed|matching our launch infrastructure to affordable and resilient space systems*. Retrieved from SpaceNews: http://spacenews.com/matching-our-launch-infrastructure-to-affordable-and-resilient-space-systems/.

UK National Space Technology Steering Group. (2014). *National space technology strategy 2014*. Retrieved from Kingston University London website: http://sec.kingston.ac.uk/uklaunch/docs/National%20Space%20Technology%20Strategy%20April%202014.pdf.

Union of Concerned Scientists. (2017). *UCS satellite database*. Retrieved from Union of Concerned Scientists website: https://www.ucsusa.org/nuclear-weapons/space-weapons/satellite-database#.Wp_odWrwa70.

United Nations. (1998). Small satellite missions. *Third United Nations conference on the exploration and peaceful uses of outer space. A/CONF. 184/BP/9*. UN.

United States Government Accountability Office. (2017). *Surplus missile motors: Sale price drives potential effects on DOD and commercial launch providers.*

Vernile, A. (2018). *The rise of private actors in the space sector*. Springer International Publishing.

Wekerle, T., Bezerra Pessoa Filho, J., Eduardo Vergueiro Loures da Costa, L., & Trabasso, L. G. (2017). Status and trends of Smallsats and their launch vehicles—An up-to-date review. *Journal of Aerospace Technology and Management, 9* (3), 269–286.

Werner, D. (2017, November 20). *Finnish startup Iceye to supply DoD unit with Airborne imagery*. Retrieved from SpaceNews: http://spacenews.com/finnish-startup-iceye-to-supply-dod-unit-with-airborne-imagery/.

Xinhuanet. (2016, December 27). *Full text: China's space activities in 2016*. Retrieved from Xinhuanet: http://www.xinhuanet.com/english/china/2016-12/27/c_135935416_2.htm.

Zak, A. (2016, July 23). *The Russian space industry at the turn of the 21 century*. Retrieved from RussianSpaceWeb website: http://www.russianspaceweb.com/centers_industry_2010s.html.

Zak, A. (2017, November 3). *Sputnik launch vehicle (8K71/8A91)*. Retrieved from RussianSpaceWeb website: http://www.russianspaceweb.com/sputnik_lv.html.

MIX
Papier aus verantwortungsvollen Quellen
Paper from responsible sources
FSC® C105338

If you have any concerns about our products,
you can contact us on
ProductSafety@springernature.com

In case Publisher is established outside the EU,
the EU authorized representative is:
Springer Nature Customer Service Center GmbH
Europaplatz 3, 69115 Heidelberg, Germany

Printed by Libri Plureos GmbH
in Hamburg, Germany